A ChatGPT Guide for Beginners

What It Does, How It Does It, and Why You Should Use It

By

Todd Bitfold

Todd Bitford

2

Disclaimer

The author and publisher of this ChatGPT guidebook have made best efforts to ensure that the information present in this book is accurate, factual, and up-to-date. However, they make no warranties (implied or otherwise) as to the accuracy or completeness of the contents herein, and cannot be held responsible for omissions, errors, or outdated material.

The information given in this guidebook may not be suitable for every individual or group and career-related effort of any particular individual and/or organization, and it is advised that anyone implementing the guidelines in this book in their day to day decision/s should do so purely at their own discretion. The publisher and author assume no responsibility whatsoever for any outcome (including litigation) resulting from the application of the information found in this book, either through following these guidelines or formulating policies based on the same.

Neither the publisher nor author shall in any way be held liable for any damages, including but not limited to special, incidental, consequential, or other forms of damage.

If anyone has any legal concerns, please talk to a qualified legal professional first (or someone with specialist knowledge of this field), before acting on the information provided in this guidebook.

Content

Content

About The Author

Todd Bitfold is a writer and researcher with a keen interest in the transformative power of new technologies for ordinary people. Todd's writing style is straightforward and accessible to readers of all levels of technical expertise, making his work a valuable resource for those curious about AI.

The goal of this book is to make AI more approachable and to foster a better understanding of its capabilities and limitations. Todd provides practical guidance on how to harness AI's power in various fields, including healthcare, finance, transportation, and education.

Chapter One: Artificial Intelligence - No Need to Be Afraid of the New Kid on the Block!

> AI's not there to take your job - but to make you much better at it!

For many people, the idea of artificial intelligence conjures images of terminators and cyborgs. However, the reality is actually quite different. In fact, artificial intelligence has made inroads across a wide variety of industries and effectively revolutionized the way things are done. Think of it as a really smart assistant... on steroids!

AI in Healthcare

Let's start with the healthcare sector. AI has made some pretty big strides in this field by helping medical researchers and doctors diagnose diseases more efficiently and with a greater degree of accuracy. Machine learning algorithms, for instance, can analyze medical images such as MRIs and X-rays and assist in detecting any abnormalities that might not be visible to the naked eye. In this way, AI helps aid disease identification at an early stage. This capability has the potential to save lives by facilitating faster intervention and subsequent treatment of patients.

Revolutionizing Transportation

Self-driving cars aren't the stuff of science fiction anymore. In fact, they are well on their way toward becoming an everyday reality. Thanks to the incredible advancement of AI technology, many self-driving vehicles can now navigate roads autonomously. This will help reduce the risk of accidents caused by human error.

It will also be a game-changer for people who can't drive because of physical disabilities or age. Technology will allow them unprecedented independence by enabling them to commute safely and easily.

The Retail Trade and the Era of Personalized Shopping

The retail industry is going through a massive transformation, largely because of AI. Let's take personalized recommendations as an example. You've probably noticed those "recommended for you" sections while browsing your favorite online shopping sites. Companies now have the capability to analyze their customers' browsing and purchase data and suggest products tailored according to their individual preferences.

It's positively uncanny how accurate the suggestions become over time. That's AI at work – right there! Think of it like having your own virtual personal shopper who happens to know precisely what you want.

Automation in Manufacturing

This is another area where artificial intelligence has had a huge impact. Many AI-powered automation systems are now increasing efficiency and productivity on the assembly lines and the factory floor. They can perform various repetitive tasks with a very high degree of speed and precision, leaving their human counterparts free to focus on the more complicated and creative aspects of manufacturing and production. Such AI-based robotics actively help improve output, even as they enhance the safety of the workplace by reducing the risk of accidents.

AI and the World of Finance

AI has become very popular in the financial world, as well, where it is used in areas like risk assessment and fraud detection. Today's cutting-edge machine learning algorithms can easily and efficiently analyze vast amounts of financial data. They can also help in the identification of anomalies and patterns that their human counterparts might miss. Apart from helping prevent fraudulent activity, AI can also make much more accurate predictions regarding the trends in the financial markets. It's like hiring a leading financial expert with an almost superhuman capability of processing and analyzing data in real-time.

Helping the Entertainment Industry Stay in Tune with Its Audience

The showbiz world has also embraced AI with a passion. If you are used to live-streaming platforms like Spotify, Amazon Prime, and Netflix, you'll see content recommended according to your tastes. AI algorithms regularly analyze your listening and viewing habits and recommend content you want to see more often.

The core aim is to enhance the end-user experience and keep people entertained. Apart from that, producers are also using AI to create highly realistic visual effects and graphics in movies and video games to ensure an even greater immersive experience for our enjoyment.

AI and Agriculture

Artificial Intelligence has also found its way into the agricultural sector. Farmers are now analyzing data from an array of satellites, sensors, and drones. This has helped them make informed decisions regarding irrigation, pest control, and overall crop management. These practices are designed to substantially reduce wastage, improve crop yields, and promote the concept of sustainable farming. AI is also being used to help predict general weather patterns to optimize planting and harvesting schedules. In a nutshell, AI is helping make farming more resilient and efficient.

Customer Service: Ushering in the Era of the Chatbot

Finally, let's not forget customer service. AI-powered chatbots are rapidly becoming more effective at simple tasks, such as handling straightforward inquiries and FAQs. They can even simulate human-like conversations, leaving the human CS personnel to handle other parts of the workload and resolve customer issues.

Instead of listening to pre-programmed music, the customer can explain their problem, and the bots will forward the relevant information to the real CS representatives. This way, they have a head start on the problem and can ensure its eventual timely resolution. Apart from reducing waiting periods, these chatbots also improve customer satisfaction by ensuring that queries are answered as soon as possible.

There you have it—a brief snapshot of AI's impact across various industries, spanning from customer service to transportation to finance and the retail trade sector. AI is slowly transforming the way things are being done everywhere.

This is no bogeyman waiting to snatch your job - but a wingman designed to make your life easier and more efficient. The possibilities in many fields are endless — and we're barely scratching the surface of what the AI revolution can achieve.

These are exciting times, indeed!

Chapter Two: ChatGPT - What's the Hype All About?

When Mira Murati, Chief Technology Officer at OpenAI, first uploaded ChatGPT, she had no idea she had upended the world. Open AI's ChatGPT gained over one million subscribers in less than a week, and within three months, the numbers rocketed to 100 million!

These are truly staggering figures and show that artificial intelligence has become a permanent part of the mainstream.

The New Wave Comes of Age

Contrary to popular belief, modern human society has advanced less than as a result of evolution than as the result of 'disruption.' The age of industrialization and its subsequent mass communication radio and TV-based era changed our collective lives within a century. The internet and other digital technologies brought their own wave of disruption. As with every innovative wave, there were those who adapted to the change and survived and those who didn't and subsequently disappeared without a trace.

Until only a few years ago, AI was seen as a fad that few people took seriously outside R&D (Research and Development) labs. Now livelihoods are on the line as people scramble to keep up.

Let's consider the example of Kodak, the first company to mass-produce cameras for hobbyists and professionals alike. They actually invented digital cameras but refused to market them, since the technology would hurt their film processing business. By the time they realized their mistake, it was too late, and the world had moved on. The most iconic camera maker in the world ended up in the dustbin of history.

This applies to every technological innovation. The lesson is that we can either manage the change and learn how to benefit from it - or become irrelevant.

Yes, ChatGPT has created a great deal of uncertainty, which is to be expected. It's a radically new and advanced technology, and we don't know where it's headed, and that's a scary thought.

As a writer, my knee-jerk reaction was roughly the same as that of horse-drawn carriage drivers when automobiles entered the market. I wanted it to go away and leave us alone! Nevertheless, over time I realized I had become too complacent. AI has made me recognize my limitations - and shown me how to move beyond them.

This software is not a cute and entertaining chatbot that you can use to while away a few boring hours. Once you learn to use it properly, it can help you write stellar essays, stories, and articles. At the same time, it will enable you to upskill yourself and become a more effective member of the workforce.

How Does It Work?

The app operates on the principle of natural language processing (NLP) technology, which enables it to understand user input and thereby generate meaningful responses. This allows the user to ask questions and receive reasonably accurate AI-based answers.

ChatGPT is trained on reams of digital information, including articles, books, and web pages. This helps ensure it generates accurate responses on diverse topics, ranging from sports and politics to science and technology, and everything in between.

You can use ChatGPT through either a web-based chatbot or a messaging app. You simply type a query on the chat interface and wait for it to respond. Its AI model uses a combination of statistical analysis, pattern recognition, and its own contextual understanding to generate realistic and human-like responses.

The app is also packed with several handy features that enable you to customize its responses and even create custom conversations. One of the main keys to its surging popularity is its intuitive and user-friendly interface that boasts a clean and uncluttered design, making navigation a breeze.

This innate simplicity is a crucial aspect of ChatGPT's tremendous success, as it allows its users to focus on the task at hand without becoming lost or overwhelmed by the technology. If you are still uncomfortable with its user interface (UI), you can personalize it according to your preferences.

ChatGPT is on the verge of revolutionizing the way machines interact with humans.

Under the hood, ChatGPT uses cutting-edge analytics that track individual user behavior to help improve the end-user experience. Since it is capable of 'learning' your preferences, it can quickly tailor its responses accordingly.

In fact, ChatGPT is unique and different from other AI models precisely because of its ability to learn from, and adapt to, user feedback. When you interact with the app, you'll have a chance to rate the quality of your interaction, and it will use these ratings to further refine itself and improve its performance.

In a sense, you are actually training it to give better answers to your individual queries. Eventually, you won't have to spend time trying to come up with targeted and relevant prompts, since the app will already know what you want.

The company's latest iteration, ChatGPT 4.0, is set to push the boundaries even further with new capabilities, features, and enhanced accuracy.

What Makes ChatGPT So Great?

Seriously, take a look at this screen:

Fig 1

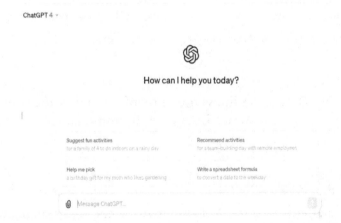

Seriously, take a long look. See any complex menus, drop-downs, options, buttons, or tabs? Instead of all that clutter, what we have is a single input field at the bottom of the screen. One that allows its users to do everything they want with this product.

It's Easy to Learn – Very Easy!

You'll be surprised at how quickly you'll be using this app like a pro. After all, it looks like any other person-to-person chat software, like Skype and Slack, and is just as easy to learn. In a way, that's what it is – software that behaves like a real person who is able and willing to help and guide you in every task it can.

From its interface to its language, it tries (quite successfully) to behave like an actual person to ensure as natural a human interaction as possible. That's why you'll be able to use it so well and so quickly.

Translation Capabilities

These capabilities allow you to use the software in any of its pre-defined languages. You can translate text and even create your own poetry and fiction. If coding is your thing, it can create step-by-step guidelines to help you, even if you're doing it for the first time. In fact, the app is capable of helping you solve many real-world practical problems.

Making Friends with This Chatty New Stranger

Since ChatGPT has been programmed to emulate the human emotional touch, it makes sense to use human analogies. Imagine you are at a party and meet a friendly and likable person. They might beguile and charm you with ideas, anecdotes, and stories to keep you entertained. But this particular stranger can offer you help, as well. Let's say you want to order pizzas for a party, and you can't figure out if three medium 9-inch pizzas would feed more people than two large 12-inch ones. ChatGPT would know.

Granted, these tools can provoke controversy and emotional response, with many people worried that they will take over their jobs and eventually society itself. They invoke the same kind of response as when Thomas Edison played "Mary Had a Little Lamb" on the world's first gramophone. Much the same aura of mystery as Edison's invention now surrounds this radical new technology. Unlike data scientists, the layman has no idea how the app is learning so fast. Many find this mystery as promising as it is tantalizing, while it has the opposite effect on others, who find it threatening and even sinister.

It Gives Advice and Support

You can use ChatGPT as a reliable source of support and advice. If you have a problem and are looking for an objective opinion, then ChatGPT can be your best friend. It can give you insights from a different perspective and even offer suggestions for whatever problem you face. While you should always carefully weigh ChatGPT's guidance with common sense, it can prove to be a valuable resource for any number of situations.

As a content creator, I have personally found ChatGPT to be very helpful in conducting research. Instead of sifting through hours of irrelevant websites, it can help me find the quotes and phrases that add that extra punch to my writing. It also helps make complicated issues easier to understand. If I want to understand something and find it too difficult, I can simply tell the app to explain it to me as if I am 12 years old, and it does precisely that.

In such cases, armed with my new knowledge, I'm able to comprehend an issue and, more importantly, help my target audience understand it as well. This is the kind of valuable result that makes the whole ChatGPT experience worthwhile.

Like me and my fellow writers, even lawyers today are learning to rely on this app to find and summarize case law on a particular topic. ChatGPT is particularly helpful in rephrasing specialized topics so anyone can understand them. It's also very useful for doctors as they prepare to speak with their patients. So you can see, this app can be useful in a wide variety of professions and in a host of ways that can make life easier for its users.

Chapter Three: Getting Started with ChatGPT Plugins and Hidden Features

ChatGPT's advanced plugins are your gateway to a more informative and interactive digital world!

OpenAI has recently released multiple plugins for ChatGPT Plus (the version of the chatbot meant for individuals) and ChatGPT Enterprise subscribers (the version intended for businesses), and the results have been revolutionary. These plugins enhance the bot's inherent functionality by allowing you to interact with its artificial intelligence in ways that were previously unimaginable.

What Are Plugins?

Plugins are also known as extensions or add-ons. They are software programs designed to enhance the functionality of a larger 'parent' application or program. Once installed, a plugin enables the parent application's users to extend and even customize the capabilities of the base software according to their own specific needs or preferences.

Some of the more common examples of plugins are described below:

Chrome

This web browser has plenty of plugins, such as ad blockers and sound boosters. These plugins (referred to as extensions in the Chrome ecosystem) offer everything from custom themes to password management.

Content Management Systems (CMS)

Platforms like WordPress enable users to add different kinds of plugins for functions like social media integration, SE (Search Engine) optimization, and even e-commerce capabilities.

There are many other plugins for audio/video software, online games, and other services. While ChatGPT was late coming to the plugin game, it has been making up for lost time.

There has been great demand for plugins to customize the ChatGPT experience and unlock its full potential. As a result, OpenAI launched ChatGPT plugins for both developers and users. It has also started to implement across-the-board support for plugins for its ChatGPT+ (ChatGPT 4.0) chatbot.

According to OpenAI, these tools have been specifically designed for its most advanced language models and have been created with safety as a core principle. They help ensure that the chatbot is capable of running computations, and that it has access to the most up-to-date information available online.

ChatGPT Plugins: Making Your Favorite AI-powered Bot Smarter!

ChatGPT 3.5 is trained on a dataset that was collected only through September 2021, so its responses are based on the world as it was until that static point in time. Training an AI model is an extensive process that requires considerable computational resources and time, so the models do not update their knowledge after training is completed. Luckily, OpenAI launched ChatGPT 4.0, the latest version of ChatGPT, on March 23, 2023. This new and upgraded version has the ability to access the internet, and it's not solely dependent on data fed into its systems through September 2021.Plugins are now being used to enhance the chatbot's current capabilities. In fact, many AI industry experts have dubbed these plugins the 'eyes and ears' of the ChatGPT world.

These plugins are designed to feed directly into the ChatGPT ecosystem. Once installed, they will give the bot access to a vast range of information and knowledge, courtesy of its third-party partners. They will help it run computations and even access current information. Thanks to these plugins, we no longer have to worry about ChatGPT's standard "Sorry, I have information only up until September 2021" apology.

As you might imagine, this update is incredibly exciting, and if used properly, it can take an already powerful AI language model to the next level.

Some of the initial beta plugin options include integration with web browsers like Chrome and Firefox. This is necessary to access new information sources and ensure that the bot provides the latest and most relevant information. Powerful code interpreters will soon be introduced for programmers and developers. These plugins also provide influencers, other content creators, and consumers the ability to create transcripts of YouTube videos.

These plugins are finally universally available to all ChatGPT Enterprise and ChatGPT Plus users.

Installing, Enabling, and Managing Plugins

Installing ChatGPT Plugins

To install plugins, log into your ChatGPT account via a web browser (this method works for both mobile and desktop systems). Alternatively, you can open the ChatGPT app on Android or iOS.

It's important to note that you will have to become a paid plan user to activate these extensions. If you have access only to the basic ChatGPT 3.5 plan, you won't be able to use plugins. But once you upgrade to a paid subscription, you'll gain access to all the new features, including third-party extensions.

Installing plugins on your ChatGPT Plus or Enterprise variant is quite simple, as we'll see.

First, start a new chat with GPT-4 by clicking on the "GPT-4" icon from the dropdown menus at the upper left hand side of your screen.

Fig 2

Now click the "Plugins" tab on the default drop-down menu. Don't worry if it says, "No plugins enabled." Simply click on the "Plugin store." This tab will grant you access to all plugins available for installation. These plugins are filtered on a popularity basis by default.

If you see a plugin you like, hit the "Install" icon under its name. Repeat this process for every plugin you want. That's it, you're done!

Take note, however, that while you can install multiple plugins, you can enable only three at one time. Plugin icons will automatically appear beneath your GPT 4.0 mode toggle. This will show you which plugins are active at any given time.

Enabling and Managing Plugins

Once you upgrade your subscription and have installed your desired plugin(s), go to user settings and click the 'three dots' icon beside your username (on the bottom left of your screen). When you have found it, do the following:

- Select "Beta features"

- Toggle on the plugin icon and enable the plugin you want

It's just that easy!

If you want to enable another plugin, you will have to deactivate at least one so you don't go over the three-plugin limit. If you want to deactivate a plugin without uninstalling it, uncheck the little box visible next to the plugin you want to deactivate. This option is available in the same dropdown menu.

The Top ChatGPT Plugins for All Your Needs

Now that you know what plugins are all about, and how to install and enable them, let's see what's out there. The plugins mentioned below include self-hosted plugins from OpenAI as well as third-party extensions. Every plugin adds its own unique ability that can be used to leverage and enhance ChatGPT's own capabilities.

Instacart

Thanks to the Instacart extension for ChatGPT Plus/Enterprise, it's now possible to order groceries from the neighborhood grocery store directly within ChatGPT itself. However, Instacart isn't just a grocery ordering application. After all, there are already plenty of those around. You can also use this plugin to enable ChatGPT to give you suggestions for recipes and meal ideas.

First, ChatGPT will provide you with the recipe and the list of ingredients, and once you have decided what you want, Instacart will enable you to shop for the products. This way, you will only have to focus on cooking … and of course eating your meal!

You can also use the Instacart extension to ask ChatGPT for dinner suggestions according to the ingredients you already have available in your kitchen. If anything is missing, Instacart will simply order it for you. All in all, a pretty useful app if you're in a hurry and want to whip up a delicious meal quickly!

Expedia

This plugin can help you coordinate every aspect of your vacation. After you install this add-on, you'll be able to treat ChatGPT like a travel agent. For instance, you can ask it to find you the most budget-friendly traveling plans, it can advise you regarding accommodations and recreational activities, and you can even have a conversation about the best flight options available, along with amenities, discounts, and recommendations. In short, installing and enabling this plugin will streamline your travel organization, making it much easier and more convenient.

PromptPerfect

PromptPerfect does exactly what its name says. It's designed to enable users to generate the perfect prompts for ChatGPT so the AI-powered bot will deliver the best possible responses. Once you install PromptPerfect, you will be able to transform the simplest prompt into a highly detailed alternative - one that has been tweaked to leverage ChatGPT's natural and descriptive language skills.

Whether you are new to ChatGPT or have plenty of experience interacting with the bot, PromptPerfect will help you create both simple and complex prompts to enhance its overall efficacy.

Klarna

This extension enables you to search and compare prices of various products from hundreds of online outlets - right from ChatGPT itself! It will help find whatever you are looking for, along with product prices, courtesy of Klarna's up-to-date database of online retailers and prices.

Apart from being a database, Klarna is also capable of utilizing ChatGPT's language processing capabilities. This means you can ask it for inspiration or even curated lists for different themes, from anniversaries to birthdays to retirement parties and almost anything else that comes to mind. If you are confused about what kind of gift to buy someone or even if you want something for yourself – ChatGPT equipped with the Klarna plugin will show you all the available options.

Chatwithpdf

This useful plugin enables you to input any PDF link on ChatGPT and receive its summary. It allows you to ask detailed questions about the content and provides accurate answers. It's like talking to an expert on the topic of the PDF. Suppose, for example, that you need to know which chemical compounds have been discussed in a scientific journal article. Chatwithpdf will tell you all you want to know, so you won't have to read the entire document.

It can even extract information about specific events. On top of all this, the extension is smart enough to condense the information present in very large PDF files to avoid exceeding ChatGPT's total word limit.

Video Insights

Video Insights is an incredibly handy plugin that will enable you to interact with multiple online video platforms like Dailymotion and YouTube. You can use it to request metadata information and transcripts from these platforms. This extension can be very useful if you see an interesting but overly long video, and you don't have time to watch the whole thing. Video Insights will go through the video, extract relevant information, and generate a short summary for you.

Stories

This extension is ideal for all you creative ChatGPT users out there. Stories can help you create illustrated stories and other documents in ChatGPT itself. If you are a blogger and want to create that 'wow factor' in your blogs, Stories has you covered. It is also useful for creating marketing collateral and other types of content. Using this plugin will help you craft the very best stories you can. It also offers SE optimization, along with multiple options for custom illustrations and fonts. Think of it as a one-stop shop.

Speak

The Speak plugin is a powerful language-learning assistant and translator, as its name suggests. Unlike other online translation apps and extensions, Speak can actually help you understand even the subtle nuances of many different languages. Are you aware, for instance, that there are several different types of polite language in Japanese, and using the wrong type can cause offense? Speak can resolve this issue for you.

It even provides Romaji, a system of writing Japanese that uses the Latin alphabet. It's a great solution for people who can't read Japanese script.

PlayListAI: Spotify Playlist Generator

You can use this plugin to auto-generate your Spotify playlist. Just ask ChatGPT to create your playlist according to your preferences – the nifty little playlist generator will handle the rest. If you are in the mood for musical gems that never hit the mainstream – say no more. PlaylistAI will spring into action via ChatGPT, and weave a symphony of diverse and eclectic sounds that will cater to your every musical whim!

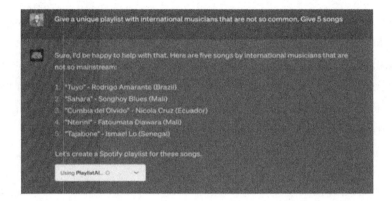

Keymate AI

This is the perfect plugin to use if you find internet searches tedious and boring. Thanks to its advanced algorithms, it will quickly and thoroughly sift through, literally, millions of data points to ensure you get the most relevant results. If you want to write a paper on theoretical cosmology, for instance, you can trust Keymate AI to take a deep dive into the heart of the topic and do all the heavy lifting for you. It's designed to make ChatGPT more accurate and efficient.

Open Table

Now there's no need to navigate through complicated apps or call anyone to make restaurant reservations. Thanks to the Open Table plugin, booking a table at your favorite restaurant is a breeze. Just ask

ChatGPT to do the job for you via this plugin. Your family and friends' dinner plans and your own date nights will definitely become smoother and more hassle-free. For example, after installing and activating this plugin you can ask ChatGPT to make a reservation for five people at nine p.m. next Sunday. It will use this extension to make the reservation on your behalf.

Link Reader

Now imagine an AI-powered bot that can read web content. The Link Reader ChatGPT plugin does exactly that. Provide ChatGPT a link and it will use this extension to scan the link and extract all relevant information. It will summarize blog posts and provide key points from articles and other online content. If you're in a hurry, this plugin is an awesome time saver. You can use it to boost your productivity while staying ahead of the curve.

These are just a few examples of different plugins and how they can automate tasks and create opportunities to make your life easier. Even better news - developers are currently working to create more advanced and handy extensions that will turbocharge ChatGPT's capabilities.

Top Hacks You Can Use to Increase Your Productivity

ChatGPT can help you increase your productivity in ways you may never have thought of before. Let's discuss a few really easy-to-use hacks:

Study Efficiently

ChatGPT is particularly useful for students who want to increase their learning speed. Let's say you pulled an all-nighter learning about the

American War of Independence for an upcoming exam. However, you aren't really sure how well you have grasped the topic. You'll be happy to hear thar you can ask ChatGPT to generate a dozen questions to test your knowledge. You can even ask it to check your answers and grade you accordingly before tackling the real exam.

Fig 3

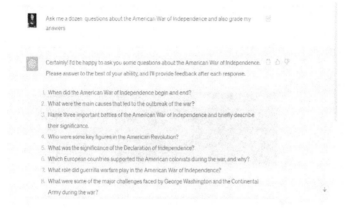

Fig 4

As you can see in the above example, not only has it provided a detailed answer but also graded my response.

Play Games or Discuss Philosophy

If you want to while away a rainy afternoon, ChatGPT will provide you all the entertainment you need. Ask it to tell you the meaning of life, and let the conversation take you into uncharted spiritual territory.

If that's not your cup of tea, you can play games with it just like you would with a human friend.

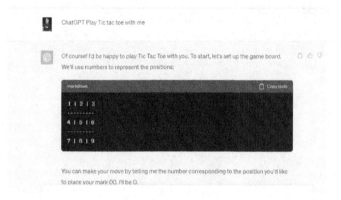

Fig 5

Read Faster

Maybe you have an important meeting and don't remember or didn't read the key points of the presentation you were given ahead of time. No need to fret - ChatGPT has your back. Just paste the documents and presentation into the chat window and ask it to highlight the most crucial points, so you will be fully informed and well-versed if asked to contribute to the discussion.

Unleashing ChatGPT's Hidden Features

Most people use ChatGPT as a writing and information collection assistant, roughly on par with search engines like Google and Bing. But as we've seen, it's a lot more than that. You can unlock its potential by using the below little-known features and capabilities.

System-level Instructions

ChatGPT has the unique ability to process directives in different modes. If you frame your prompt as a specific system instruction, you will be able to guide its behavior in a particular manner. For instance:

"ChatGPT, You are an assistant from the Tudor age and you speak like William Shakespeare."

These instructions will ensure that ChatGPT will generate responses in the linguistic form used during late medieval and early renaissance eras in England.

Iterative Refinement

You'll be happy to know that you don't have to settle for the very first answer generated by ChatGPT. In fact, the initial response might simply provide you with a specific direction, which you then use to further refine your question or ask for clarification. The iterative refinement approach usually leads to information that is more relevant and precise.

Example

You: *"What can you tell me about quantum mechanics?"*

ChatGPT: *"Quantum mechanics is a specific branch of physics..."*

You: *"I want you to explain the concept in as simple terms as possible."*

ChatGPT: *"Certainly! Just imagine that the planets and galaxies work as dice being rolled..."*

Combine Multiple Instructions

It's possible to combine several instructions if you want to achieve a specific behavior:

"ChatGPT, Tell me how to lose weight while being a pirate who only speaks in rhymes."

ChatGPT will adopt a pirate persona while giving you advice on how to lose weight – and will do so in verse!

It also appreciates being appreciated!

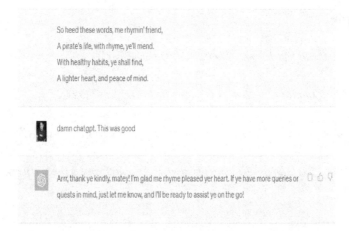

Situational Behavior

It's even possible to set up a specific situation with regard to how the model should behave throughout your interaction:

"ChatGPT, You are a detective trying to solve a mystery."

ChatGPT will respond as if it is actually investigating a case. It might even try to uncover clues and ask probing questions.

As you learn to use ChatGPT, you'll quickly realize that it learns along with you and will better understand your questions and requirements more as time passes. With the right features and plugins, ChatGPT can help you maximize your leisure and work hours.

Chapter Four: Using ChatGPT for Personal Finance

ChatGPT's ability to offer financial advice and information is one of its coolest and most useful aspects. From calculating taxes to managing your expenses, ChatGPT can help you in a multitude of ways. It can also help you find new ways to invest your savings based on your risk tolerance, time horizon, and financial goals.

You will, of course, have to list all your expenses, income, debt, and explain your financial requirements before you start using this AI model for investment purposes. Once it has the required information regarding your investment and savings habits, it will advise you accordingly. If utilized correctly, it will help you achieve all your short- and long-term financial objectives by sharing its comments on your specific investment and savings habits. It may advise investments, but obviously it is your decision to determine if this advice is useful to you, and will benefit you in the long run.

Understanding the Basics for Personal Finance

Before you start asking ChatGPT for financial advice, it's a good idea to understand the basic principles of personal finance. Some of the fundamentals include:

Budgeting

A budget is a simple plan that outlines your expenses and income. You have to calculate your total annual income and keep track of your expenses every month. This will help you understand where your funds are going. A budget will help you allocate your income to different categories, like transportation, housing, savings, groceries, and debt repayment. While ChatGPT can give you budgeting advice, it cannot maintain your budget for you.

You will have to figure out your expenses yourself. If you find it's too much of a hassle, you can use a good budget app, such as YNAB, Mint, or Pocket Guard. Once you know how and where you are spending your money, you can ask ChatGPT to help you curb your spending and ensure you're not spending more than you should.

Savings

Saving money is vital for achieving financial goals, building an emergency fund, and preparing for the future. It's always a good idea to save a certain portion of your income as regularly as you can. You may start with a relatively small percentage and increase the monthly amount over time.

Debt Management

Managing debt is imperative for long-term financial well-being. You should prioritize paying off your high-interest debts, such as credit cards and short-term loans, as soon as possible. ChatGPT can help you create a debt repayment plan to ensure you make consistent payments and reduce your overall debt burden. However, you will have to stop spending beyond your means and use your credit lines responsibly.

Emergency Funds

Emergency funds act as a safety net for sudden and unforeseen expenses like home repairs, car repairs, medical bills, or job loss. You should try to have at least three months' worth of living expenses in your emergency fund. You should also consider keeping these funds in an easily accessible account so you can withdraw some or all if necessary, but do your best to replace anything you withdraw.

Insurance Coverage

Insurance plans are crucial for protecting yourself, your loved ones, and your assets against natural disasters and other calamities. Some of the more common types of insurance include auto insurance, home/renters insurance, health insurance, and life insurance. You should review your insurance coverage at least once a year to ensure it meets your requirements. You can also ask ChatGPT for advice regarding insurance plans.

Financial Education

ChatGPT can be very useful here, since it will aid you in continuously educating yourself about personal finance to ensure you always make informed decisions. Just take care to also educate yourself regarding the world of finance in general. Read books, attend seminars and workshops, and visit reputable financial websites. Once you have the required knowledge, you can use it to create the right prompts for your ChatGPT financial advisor.

Setting Your Financial Goals

ChatGPT can help you clarify your objectives and set your financial goals. In fact, if used correctly, it can also create your own personal roadmap for a stable financial future. Let's see how you can leverage ChatGPT to set your individual financial goals:

Define Your Goals

Begin by clearly articulating and defining what you want to achieve financially. In other words, where you would like to be with your finances in a year, two years, or even five years. Maybe you want to start a business, build an emergency fund, save for a new home, pay off debt, or even retire to a comfortable post-work life. Here, you must be specific about the funds you need to achieve your objectives within the expected timeframe.

Conduct Research and Gather Information

Since it's your financial future at stake, you must ensure due diligence in your data collection efforts. Ask ChatGPT about different financial topics related to your objectives. You can, for instance, inquire about budgeting techniques, investment strategies, tax planning, debt repayment strategies, and any other relevant topics. The AI model will help you generate insights and educate you on the myriad aspects of personal finance that are relevant to your particular goals.

Evaluate Your Existing Financial Situation

Be sure to provide ChatGPT with information regarding your current financial position, such as your annual income, liabilities, expenses, assets, investments and savings. This will help it understand your monetary situation and enable it to give you valuable advice about what adjustments might be necessary to reach your objectives.

Ask for Specific Advice

It is possible to seek personalized advice from ChatGPT according to your goals and based on your current status. You can, for example, ask the model about the best investment and savings options for your desired timeline. It can also give you advice regarding strategies for increasing income and reducing expenses, and tips for creating an effective budget. Apart from that, you can directly ask it to help you prioritize your financial goals.

Create a Financial Game Plan

You should be able to develop a financial plan based on ChatGPT's data and advice. Your plan should ideally outline the specific steps you need to take to achieve all your financial objectives. This plan has to include actionable points, such as paying off debts in order of priority, investing in specific assets, or saving a specific amount each month.

Set Milestones and Track the Progress of Individual Milestones

Break down your main plan into smaller milestones, or targets. Monthly targets are often best, since you can easily modify them every few weeks, or as and when required. You can also opt for quarterly, bi-annual, or yearly checkpoints to measure your progress. However, you must regularly review your targets and track your progress. This will help you stay motivated and be in a position to make adjustments for unexpected circumstances.

Adapt and Learn Continuously

ChatGPT can be your companion and advisor throughout your financial journey. The world of high finance is always changing, with new challenges and opportunities constantly arising. Being curious and well-informed can help ensure your long-term financial survival.

Seek Professional Advice

While it is certainly true that ChatGPT can provide plenty of helpful insights and advice, you have to understand its limits. It would be a great idea to consult a financial advisor and run ChatGPT's advice by these professionals. This will help you determine the AI model's capabilities - and limitations.

Key Advantages of Using ChatGPT for Personal Finance

> **Eventually, both you and your trusty AI sidekick will evolve enough for you to use its advice on a standalone basis!**

The nifty little AI model comes with immense benefits when it's used to create a well-rounded financial strategy.

Easily Accessible and Convenient

Unlike an investment analyst, ChatGPT can be accessed 24/7 on the web and via your phone app. As long as you have a stable internet connection, you can ask the AI model for critical financial advice and you'll receive instant results. This accessibility enables you to obtain personalized financial advice whenever you need it without having to schedule an appointment and travel to an investment advisor's office. Ultimately, you will be empowered enough to make snap decisions and take advantage of even the most fleeting opportunities.

Helps You Make Better Investment Decisions

Investing your money is the best way of securing a secondary income and helping your funds grow over time. With the right prompts, ChatGPT can teach you about different investment options, such as mutual funds, bonds, commodities, precious metals, stocks, and real estate. It will give you detailed guidelines regarding low-risk, low return and high-risk, high return investments.

You can calibrate its responses through trial and error until it offers options that are tailor-made for your level of risk tolerance. It will also guide you regarding the time horizon of different types of investments while keeping your core financial goals in mind. Ultimately, it will help you choose the investments that are right for you.

Can Help You Plan Your Retirement

It's never too early to start planning for retirement. Now, thanks to ChatGPT, you can actually get a head start on your retirement plans! As long as you give it the correct prompts and ask relevant questions, it will advise you on your options and suggest the best investments for your retirement nest egg.

Exceptional Knowledge and Expertise

ChatGPT has access to and has been trained on massive volumes of data, including financial strategies, concepts, and standard best practices. It can provide highly accurate and reliable information on a broad array of personal finance topics, including investing, budgeting, saving, and debt management. You can access and leverage the model's vast knowledge to make highly informed financial decisions, even as you gain insight into complex financial concepts.

Offers Personalized Assistance

This AI model can and does offer personalized assistance, since it understands and responds to individual questions, queries, and circumstances. It will provide customized suggestions and take into account your unique circumstances, preferences, likes, and dislikes.

Potent Educational Tool

ChatGPT is arguably the best educational tool for personal finance. You can learn all the terminology and concepts simply by having an interactive discussion with it, during which you ask questions and it provides answers. This will allow you to acquire knowledge at your own pace.

An Extremely Cost-Effective Solution

Utilizing ChatGPT instead of engaging the services of an investment firm can prove to be very cost-effective. It is also a lot less expensive than taking a higher education course in finance! You can use it to gain critical financial insights at no additional cost.

Providing Input

ChatGPT is only as good as the input and prompts it's fed. The level of precision and detail of the information you share with this tool plays a critical role in its suggestions. Based on your prompts, it will analyze and assimilate the information and only then decide on the correct course of action:

Prompt One

Let's start with a basic financial prompt:

"Tell me how I can invest $500 over a 12-month period."

In the above example, an amount and a timeline have been provided. This is sufficient information for the AI model to suggest a few tentative lines of action. Based on the above criteria, it will give you the following options:

High-Yield Savings Account (HYSA)

It might ask you to consider putting your money into a high-yield savings account. Such accounts may not provide high returns, but your funds will be safe and easily accessible.

Term Certificates of Deposit (CDs)

These certificates typically offer higher interest rates than regular savings accounts. However, they are time-barred, meaning your liquid funds may be blocked for the duration of the stipulated term, which could be anywhere from one month to many years.

Stock Market

Investing in individual stocks can be very risky, especially if you have a small sum to begin with. However, you might consider low-cost brokerage platforms. These platforms offer fractional share investments. This way, you will be able to purchase portions of a stock rather than a full share. It is crucial to conduct in-depth research on the shares you are interested in before making any investment decisions.

Potential returns on a mere $500 investment for one year won't be very high, so it's important to set realistic expectations for your returns.

As always, it's best to consult with a financial advisor, and if that's not possible, conduct your own research to find the best investment options, keeping in mind your financial goals and risk tolerance.

The above answers are based on their relevance to your prompt. If you'd like more options with the same prompt, you can always ask ChatGPT to regenerate its response.

Now let's change the prompt and give it different parameters:

Prompt 2

"I have $15,000 I want to invest for my retirement: Give suggestions."

With this prompt, you can expect to receive information on Individual Retirement Accounts (IRA), mutual funds, bonds, real estate, the stock exchange, etc. It will also offer you advice on the diversification of your portfolio.

Interpreting and Evaluating Responses

It is critical to understand that ChatGPT is an AI language model, not a certified financial advisor. While it does provide actionable information and suggestions, it's better to take the opinion of professional financial advisors before making a major investment. Here are a few guidelines that will help you evaluate ChatGPT's responses:

Understand Its Limitations

ChatGPT's responses are based on information it has in its database. This means it probably won't have regular access to the latest regulations, financial news, or the kind of personalized information it will need to advise you on your specific financial situation.

Look for Consensus

If ChatGPT consistently regenerates the same suggestions and recommendations, this might indicate that it has a better understanding of best practices. Ultimately, it means safer and more productive results.

Verify Critical Information

Always independently double-check facts and financial information you receive from ChatGPT. You should cross-reference the information with multiple reliable sources, including reputable financial websites and publications to authenticate the AI model's recommendations.

Exercise Prudence

Remember, it's your money on the line, so exercise caution when interpreting ChatGPT's responses. You have to consider your specific situation in terms of your goals, your funds, and your risk tolerance. Opt for the least risky alternative - at least in your initial investments.

Creating a Personal Budget

Creating a budget is an absolute breeze now, thanks to ChatGPT. It will help you in the following ways:

Budgeting Basics

ChatGPT can easily explain the core fundamentals of budgeting, such as expenses, income, savings, and debt. It can also help you understand the necessity of budgeting and how a well-thought-out budget can positively impact your overall financial situation.

Setting Financial Goals for Your Budget

The AI model can also help you define your financial priorities. If you want to save, pay off debts, make a specific purchase, or even build an emergency fund, ChatGPT can provide you the guidance you need.

Creating a Comprehensive Budget Plan

The information you provide ChatGPT can help you develop a comprehensive budget plan. It can assist you in allocating your income to different expense categories in order of your preference. It can also assist you in setting spending limits.

Tracking Income and Expenses

ChatGPT can recommend multiple budgeting tools and apps to enable you to monitor your income and track your expenses. If you want, it can also suggest methods for staying accountable for your expenses. You can ask it to recommend dedicated budgeting apps as well.

Expense Analysis

ChatGPT can analyze your day-to-day spending patterns to help identify trouble areas where you may be prone to overspending. After reviewing your expenses, it will suggest potential cost-cutting or finding alternatives that align with your budgetary plans and financial goals.

Saving Tips and Strategies

The AI model can provide many helpful tips to ensure you don't overspend and instead stay well within your budget. It will help you regularly review your savings goals and also track your progress. Using reminder software, you can ask it to remind you to celebrate financial milestones as soon as you achieve them. This will keep you motivated and ensure that your financial objectives are always in sight.

Adjustments and Optimization

ChatGPT can respond to changes in your circumstances by helping you adjust your budget. If you encounter unexpected expenses or experience a change in income, you won't need to reprioritize your goals because the AI model will guide you on adjusting and adapting your budget accordingly.

ChatGPT and Debt Consolidation Advice

ChatGPT can give you plenty of debt consolidation advice and help you live a debt-free life. By following the AI model's guidance, you will not only demonstrate better fiscal discipline but can also rid yourself of debt. ChatGPT can help you consolidate your debt in the following ways:

Provide Guidance Regarding the Debt Consolidation Process

Debt consolidation entails combining multiple debts into one large single loan or credit structure. ChatGPT will explain the concept as simply as possible. It will teach you to understand your debt options and consolidate them to simplify repayment and reduce your monthly interest payments.

Evaluate Your Financial Position

ChatGPT will assess all your debts and their various interest rates, as well as their corresponding monthly payments. Provided you have given it all the required information, it will help you keep track of any late fees or other penalties and outstanding balances.

Research Your Debt Consolidation Options with You

You can use several methods to consolidate debts, including home equity loans, balance transfers, credit cards, and personal loans. The debt snowball method, for example, advocates paying off high-interest smaller debts before bigger ones. Apart from that, there are debt management programs offered by credit counseling agencies. You can use ChatGPT to research the pros and cons of every option by comparing their terms and conditions, interest rates, and fees.

Consider Your Creditworthiness

Lenders always consider your credit score and associated financial history before they approve credit or consolidate your loans. Low credit scores make it very challenging to obtain favorable interest rates and easy repayment terms. Here too, ChatGPT will help you explore suitable options that align with your current credit score.

However, as I've stressed with regard to all other financial aspects, it is crucial to double-check ChatGPT's debt-related suggestions by involving qualified and experienced financial advisors. You should also regenerate responses to see which one is repeated the most. Once it's approved by a financial expert, you can feel comfortable proceeding with your financial planning and debt removal strategies. Once the experts have vetted ChatGPT's advice, you won't need to go back to them time and again; instead you can rely on your little AI chatbot for financial advice.

Chapter Five: ChatGPT for Business Growth

Few technologies can supercharge your business growth as much as AI-driven ChatGPT. By providing ideas and increasing the pace of the workflow, it can get things done faster than ever before. It's very simple, really. The model's usage is limited only by the needs of your business - and your imagination and familiarity with the tool. You can use it to grow your business in countless ways.

Enhancing Customer Support

You can use ChatGPT as a virtual assistant that will provide 24/7 customer support. It has an intrinsic ability to understand natural language. This means it can speedily address customer queries and offer solutions. In fact, it can even troubleshoot some of the common issues associated with your industry. Since customers would have their queries answered instantaneously in real-time - they will feel valued and think more highly of your brand.

Lead Generation

ChatCPT can not only collect lead information but also qualify prospective customers and make it easy for them to enter the marketing funnel. It can do this by providing them with tailored information according to their browsing history, thus setting the stage for higher and better conversion rates.

ChatGPT Can Actually Personalize Your Recommendations!

> **It's possible to integrate ChatGPT into your social media channels and website for enhancing the engagement levels of potential customers.**

You can now implement ChatGPT to offer service recommendations or personalized products to your customers. Since it can analyze their purchase history and preferences, it's possible to suggest more relevant offerings based on what they like. In the long run, it will lead to increased sales and greater customer satisfaction.

Content Creation

ChatGPT can effortlessly generate high-quality, high-value content for all your email newsletters, blogs, and social media posts. Such outsourcing can save you a ton of money, effort, and time. It will also ensure a steady flow of relevant and engaging content that will keep your target audience interested and constantly coming back for more.

Market Research and Insight

Consider tapping into ChatGPT's unique ability to analyze customer feedback. It can understand market sentiments, fads, and trends, and use the data to help you make informed decisions that will enable you to stay ahead of the competition.

Todd Bitford

User Tutorials and Onboarding

You can train ChatGPT to help new customers become acquainted with your software platform. It can help guide them through the onboarding process by offering step-by-step instructions and freeing your staff for more important work.

Social Media Engagement

It's possible to leverage ChatGPT to respond to messages, comments, and mentions on various social media posts. Instead of coming up with responses on your own, you can simply copy the post on ChatGPT and ask it for replies. You can use the regenerate option to create multiple replies and fully participate in the thread. Its personalized responses can actively help boost your brand's reputation and foster a sense of community.

Employee Training and Support

ChatGPT can help you empower your workforce. It can provide training materials and also answer many internal queries that your busy managers may have overlooked in orientation sessions. It can also act as your company's internal knowledge base, so employees are able to access information quickly and easily.

Since new employees are often reluctant to approach their superiors with questions, they could miss out on details that would help them become more productive and efficient. ChatGPT can help break down that barrier and enable them to access the information they need so they can feel confident in contributing to the growth of the organization to the best of their abilities.

Language Localization

This AI model can help you expand your business to new markets by translating and adapting your content for different languages and cultures. This will allow you to connect to audiences anywhere in the world. Since it can literally speak their language and dialect, they will welcome your company with open arms.

Fig 4

In the above example, I've used a query asking ChatGPT about its major advantages for business entities – but as you can see, I wanted the answer in Spanish. It obliged immediately.

Take note, you have to specifically tell it to answer in a particular language; otherwise, it will answer and numerate everything in its default English.

AI-Powered Business Opportunities

When it comes to business prospects and AI, there's a whole new world of possibilities! Think of ChatGPT as a sort of super-charged engine that will propel your business to new heights of success. Some of the fantastic opportunities driven by AI include:

Awesome Customer Service

Thanks to AI-based super-smart chatbots, you can now have a virtual assistant available at all hours of the day and night. Now customers won't have to wait and listen to terrible music they don't want to hear, but rather, will instantly receive the help they need.

Bespoke Marketing Strategies

Artificial intelligence enables you to take a deep dive into the world of customer data analysis. It will help you to better understand the preferences and behaviors of your target audience. That means you can create highly personalized marketing campaigns that really hit the sweet spot in terms of connecting with your audience.

Fig 6

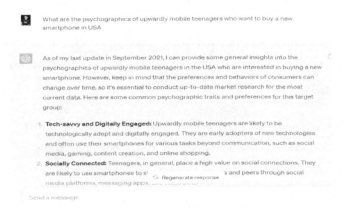

In Fig 6, we can see that it's possible to gauge an idea about the psychographics (attitudes, aspirations, and other psychological traits) of a target market - in this case, upwardly mobile teens who want new smartphones.

In other words, if you want to market a smartphone, ChatGPT can tell you the kind of specifications and pricing requirements desired by your audience. While it may not be capable of predicting the future, it can nonetheless analyze information and give you an idea about future market trends. It can also assess customer demands as well as the potential risks associated with any new business venture. Duly armed with this knowledge, you will be able to make savvy decisions for your business.

Supercharging Manufacturing Facilities

AI can predict regular maintenance needs and optimize manufacturing processes. In a nutshell, it can effectively ensure that everything runs like clockwork. Apart from increasing efficiency, it can also massively boost productivity.

Identifying the Top AI-Driven Business Opportunities

Aside from regular business prospects, AI has created several cutting-edge fields that are revolutionizing the world as we know it today:

Robotics and Automation

AI-driven robotics can be applied in warehousing, manufacturing, and many other industries that require the automation of repetitive tasks to increase productivity.

HR and Recruitment

AI can also streamline the recruitment process by identifying suitable candidates, analyzing resumes, and even conducting initial screening interviews.

Gaming

AI can be used to create highly entertaining and intelligent NPCs (non-player characters) in video games. This can help enhance the gaming experience and make virtual worlds more realistic and immersive.

Energy Management

AI can go a long way toward optimizing energy consumption at your plant or any other place of business. It can predict equipment failures, and enhance overall energy efficiency in commercial and industrial settings.

AI-Enabled Goods and Services

Today, plenty of goods and services can be enhanced by AI to perform better and more effectively. In fact, using AI to augment various products and services is one of the best ways of improving both. Some of the more common applications include:

The Smart Home

From detecting prowlers to taking pictures of anyone trying to gain access to your home, AI can do it all. It can even send the picture to local police and request urgent response. Today's state-of-the-art AI-powered smart home devices possess the capability of learning from user behavior. They can optimize energy usage, enhance security, and provide added convenience using their voice-activated controls.

Fraud Detection

AI-based algorithms can analyze transaction data and methodically check it to detect and prevent fraudulent activities in a broad range of financial services. It can check online transactions in real-time and catch criminals red-handed. AI fraud detection systems can be integrated into existing infrastructure to actively monitor transactions and related activities for signs of potential fraud.

Language Translation

AI-powered language translation services can accurately translate speech and text between multiple languages in real time. They can also interpret the subtle nuances of different dialects and guide you regarding different cultures.

Using AI Process Automation

Applying AI in process automation can significantly improve efficiency and decision-making in many industries. Here are a few steps that can help you automate your business:

Identify the Process

First, you should identify the particular process that you want to automate. It could be anything from document processing to data entry or customer support, and even predictive maintenance.

Data Collection and Preparation

Remember, AI models require copious amounts of relevant data before they can understand their tasks. Gather the required data from your existing processes and ensure that you've organized everything in sequence - just like this list!

Select the Right Technology

Select the AI tech that works best for your processes. You can utilize different technologies like natural language processing (NLP), machine learning or even a combination of the two.

Integrate Your Selection with Existing Systems

Integrate your desired AI model with your current process and check to make sure they work seamlessly together.

Testing and Validation Phase

Thoroughly test your new AI-powered process automation in a controlled environment. Verify that it functions as expected and meets all your requirements.

Deployment and Monitoring

Once the AI automation has been validated, deploy it into your live environment. You should actively monitor its performance at this stage to identify any issues and make incremental improvements.

User Training and Acceptance

Conduct change management courses to ensure the acceptance of your new systems by your employees. Of course you will also have to train your personnel to operate the new AI-powered systems.

Create a Feedback Loop

Encourage feedback from all users and stakeholders for continual improvement of the system. Then incorporate the feedback to tweak the AI mode for real-world performance.

Privacy and Security

Take note of the security and privacy implications of using an AI model for automating your processes. Take care to ensure that all sensitive data is handled safely and securely and is compliant with all relevant regulations.

Successful implementation of AI in process automation requires careful planning and seamless collaboration between different teams. Aim to start with small and manageable projects and gradually expand your AI initiatives as you gain more experience and confidence.

Utilizing ChatGPT for Enhancing the Customer Experience

ChatGPT is primarily a chatbot, and customer service is the core application for most chatbots. By handling the most common queries, it will ensure your CS representatives can devote their time to more critical questions. You can integrate ChatGPT into your automated chatbot systems to interact with your customers in real time. Here's how you can use this bot to create a positive customer experience:

Personalization is the Key!

According to Dale Carnegie's *How to Win Friends and Influence People*, most people love to be addressed by name. It makes them feel special and heard. Provided with the right prompts, ChatGPT can address individuals by their name, help create a sense of familiarity, and build trust.

Quick and Accurate Responses

Obviously, all customers appreciate fast resolutions to their problems. ChatGPT won't ask your customers to hold while it looks for answers. On the contrary, it will promptly provide relevant and accurate answers.

Seamless Integration

Consider integrating ChatGPT into your existing customer support channels on your website and your mobile applications. An across-the-board integration will provide a consistent and unified brand experience.

Use It to Gather Feedback

ChatGPT is an invaluable tool for gathering feedback from customers regarding their experience. It can ask simple questions or open-ended queries that will help you gain insights to make further improvements.

Emulate Your Brand's Personality

It's possible to tailor ChatGPT's responses to reflect your brand's tone and personality. If you've chosen an informal and friendly persona for your brand, you can teach ChatGPT to adopt the same tone in order to reinforce your brand identity.

Leveraging ChatGPT for Business Startups

If you are thinking of starting your own business, ChatGPT can help you gain critical insights. Apart from using it for content creation and customer support, it can be utilized in the following ways:

Market Research Buddy

Starting a business without understanding your target market is much like sailing a ship without a map. But ChatGPT can become your trusty companion as it helps you conduct market research beforehand. Ask it about customer preferences, industry trends, or even your competition. It will give you the valuable insights you need to make informed decisions.

Brainstorming Partner

The AI bot is an excellent brainstorming partner. Bounce your ideas off of it, and it will explore possibilities, offer suggestions, and even come up with innovative solutions.

It Will Help You Understand Your Audience

Understanding your target market and tailoring your offerings accordingly is a surefire recipe for success. ChatGPT can analyze customer preferences to provide you with personalized product or service recommendations, leading to an enhanced customer experience.

Business Strategy Advisor

Interested in scaling up your business? ChatGPT can spot market trends and analyze data and past performance to help you make data-driven decisions and devise effective business strategies.

Chatbot Integration

Integrating ChatGPT into your website is an awesome way of kick-starting your business. The chatbot will answer FAQs, engage visitors, and guide potential customers through the sales funnel. Its ability to provide a human-like touch can make interactions more productive and pleasant.

ChatGPT is an amazing tool to increase productivity and better understand your target audience. The AI bot will not only attract potential customers; it will help them every step of the way as they enter the marketing funnel – all the way to conversion.

Chapter Six: ChatGPT for Content Creation

Now that you understand how ChatGPT works, let's talk about creating amazing content and managing your time like a boss!

If you want to create content that people will love to consume – you need a game plan. You need to determine what you want to write about and create a clear outline in your mind. This will help you stay focused and organized throughout the brainstorming process.

To do this, managing your time as effectively as possible is critically important. Let's face it, we all have a thousand things to think about once we start working. If you want your creative juices to keep flowing, you need to set specific goals and clear-cut deadlines for every section of your content.

In my case, before writing this book, I first created an outline, and then broke it down into individual chapters, and each chapter into specific sections. Once I start working, I gave myself a time limit for completing my research, creating a draft, fine-tuning it, running it on editing software, and finally proofreading the finished sections.

By concentrating on individual tasks, not only am I able to create great content, I actually enjoy doing it! If you follow this system, you'll be less likely to become sidetracked, start surfing the web and otherwise end up spending too much time on one section.

But hey, we all hit that ominous writer's block sometimes, right? If you also feel like you are facing a blank wall or feeling uninspired, you don't need to fret about it - too much. It's okay to pick up your phone, run through your social media or take time out for a cup of coffee or even listen to some soothing music. In fact, you should include all such activities in your time limit.

If nothing works, talk to a friend. Sometimes, a fresh perspective is all you need to get back into your creative zone. You might be surprised at how often inspiration strikes when you are just about ready to give up.

Ultimately, creating excellent content while managing your time is about finding what works best for you. So, don't be afraid to experiment – as long as you stay organized and give yourself some grace along the way. With ChatGPT by your side, you already have the perfect tool to help you create amazing content – in record time.

You got this!

Creating Excellent Content and Managing Time Effectively with ChatGPT

Creating awesome content has never been easier - thanks to ChatGPT. However, there are certain steps you have to take before you can start using this wonderful tool. You'll have to sign up on the OpenAI website (You'll recall that OpenAI is ChatGPT's parent company). The signup process is super easy. Just use your Google ID or any other email address and create a password. That's it, you're done! After that, you can log in using your credentials whenever you want.

Before diving right in, though, you should brainstorm ideas and outline the important points. Then gather as much relevant information as you can. This will give you a rock-solid foundation to work with.

Next, you should leverage ChatGPT's formidable capabilities. After all, it is an excellent tool for generating ideas in an easy-to-understand format. It will help you refine your writing and even aid in fact-checking your data. If you find you can't seem to go on, you can also throw a few prompts at ChatGPT and see how it uses them to generate fresh concepts, perspectives, and insights.

Application of ChatGPT in Content Creation

If you use it properly, ChatGPT can become the ultimate game-changer in content creation. You can use it in many different situations, such as:

Generating Ideas

If you're feeling stuck and can't continue, it's often a good idea to talk to a friend and bounce your ideas around. Well, ChatGPT can be a great sidekick that will act as your very own idea generator. Just ask a few questions on a topic or two and see it work its magic as it takes over your brainstorming duties.

Refining Your Writing

Need help polishing your content and making it more readable? ChatGPT has your back. You can ask it to suggest improvements and even help you rephrase your sentences for maximum impact. It can also make your content more engaging and just plain fun to read.

Beef Things Up!

This AI bot can help you expand your content and make it better than ever before. You simply need to provide a prompt or a brief summary, and it will generate the required additional details, examples, or explanations. You can even create a template of its responses based on your prompts:

Fig 7

You can save the above template on MS Word or any other word-processing software you use. Once you have generated sufficient responses, you may select the ones you need to flesh out your files.

Keep in mind, generating responses can become a regular rabbit hole as ChatGPT tries to grasp your exact issue. Here are a few key points that will help you along:

Understand Your Content

Begin by reviewing all your existing content, along with all its points and sub-points. Then clearly identify the sections you want to flesh out and determine the key areas you want to explore or further elaborate.

Prepare the Relevant Prompts

This is the single most important aspect of your ChatGPT experience. Once you have learned the science and art of creating relevant prompts, your work will be half-done! Ideally, your prompts should outline specific information or a particular direction you want your content model to follow. It would be a good idea to include not just the relevant context, but also any constraints to help generate the desired responses.

Start Using Your Prompts

Type the prompt in the ChatGPT chat box model and ask it to generate additional content based on your input.

Example I

Let's suppose you're a teacher who wants to prepare a lecture on American political figures of the last 50 years for your 8th-grade students. You have around 1500 words and would like another 500 to wrap it up. You might use the following prompt:

"Write a 400 word essay on key American political figures of the past half century," *using informal English suitable for 12-year-olds.*

This prompt is both concise and specific. It has a subject matter and explains how that subject matter has to be presented and for which audience. You can add the remaining 100 words while fine-tuning the 400-word essay to integrate it into your original content.

Refine Your Prompts

Be sure to review ChatGPT's responses and assess their quality and overall relevance to your content. If necessary, keep modifying the prompt and saving the results . You can also ask clarifying questions if you are not sure of the response. Apart from that, you may also ask the model to focus on the more important and specific aspects. This 'trial and elimination' process will greatly help improve the generated output - until you end up with exactly what you wanted.

Merge ChatGPT's Content with Your Original Text

> **Always remember ChatGPT is your helper -not your substitute!**

Simply copy/pasting the generated responses won't work, as mentioned in Example 1. You have to ensure that everything aligns and gels with the overall style, tone, and structure of your original piece. Edit and revise the content as many times as required to maintain flow, coherence, and consistency.

Proofread and Refine the Finished Work

Carefully read your expanded content and also make any adjustments required to improve your piece's readability, lucidity, and overall quality. You should pay extra attention to style, grammar, and accuracy. You might consider running your work on Grammarly and Google Editor to weed out any typos as well as syntax and grammatical errors.

Maintain a Critical Eye

While ChatGPT can be extremely helpful in generating the content required to expand your work, it's still very important to critically evaluate its output. Be diligent about exercising your editorial judgment prior to finalizing any document. The model is known to produce incorrect information on occasion. Since its answers depend on its database, it's your responsibility to check all the information and ensure the authenticity and accuracy of the expanded content. This is also another reason why ChatGPT, and AI in general, won't be taking over the world anytime soon!

This is a good place to remind you that ChatGPT is basically a language model, and it might not be possible to always produce flawless output. Sure, it's a really powerful tool to assist in your content expansion efforts, but your personal oversight and judgment are essential for creating perfect content. You must ensure the editing and refinement process of your document to arrive at a suitable final draft.

Optimizing Time Management

If you are working on a really big project, such as a 100k-word book, for instance, it's a great idea to generate a table of contents. This table can help you create a rough outline of everything you want to include in your book. Once your TOC is complete, you'll need to break down the individual chapters into sections, each with their own bullet points. The sequence would go something like this:

- Core goals

- Individual chapters

- Sub-headings

It Will Help You Prioritize Wisely

You can discuss your tasks with ChatGPT, and receive suggestions on how you can prioritize them to ensure the more important sections and topics are covered first. The AI will help you analyze your preferences. This is a good way to keep on top of your daily to-do list.

ChatGPT can assist you in creating a schedule so you're able to 'timebox' your tasks. The timeboxing concept is pretty simple, and designed to ensure that you're able to maximize productivity within a given time span. You can use this technique whenever you assign a specific deadline to a task. The deadline becomes your 'timebox' and once it's finished you can analyze your progress before moving on to the next task. If the box is too big, you might need to tighten the timelines and vice versa. This is known as a timebox.

Once you wrap up your task, check your progress and move on to the next one. This way, you'll better manage your time and energy, and avoid spending too much of both on a single task while ignoring others. ChatGPT can be extremely helpful in creating individual timeboxes based on the topics and word count per topic.

On a personal level, I have found that it has increased my productivity a great deal as I write this book.

Reminders and Notifications

ChatGPT is also capable of sending you reminders and notifications regarding your upcoming deadlines or any other scheduled activity. You won't ever have to miss a crucial task and can stay on track all the time. ChatGPT doesn't have any reminder plugins on its own, but you can ask it to develop the wording of your notifications and use them with any reminder software app, such as Microsoft's To-Do or Desktop Reminder, or Google Keep, and many others.

Analyze Your Productivity Levels

If you feel your productivity is not up to speed, you can always initiate a conversation with ChatGPT. In fact, you can ask it to analyze your work, and see where you might be spending too much or (too little) time. It will provide helpful insights and suggestions on how you can optimize your workflow.

Leveraging ChatGPT for Great Content Creation

Once again, your objective is crucial if you want excellent results. That's why before you dive into a conversation with your trusty ChatGPT sidekick, you must have a clear idea of the direction of your content and your target audience. You have to know the core function of your content. Are you trying to educate, entertain, or persuade your audience? Knowing your objective will help ChatGPT guide the conversation in the direction you want.

Have a Friendly Chat

Treat ChatGPT like the friendly companion it is, and you will see instant results. It will help to start the conversation in an informal manner while using a casual tone. You will be more at ease once you establish a relaxed atmosphere. This will, in turn, encourage the model to respond to your queries in a conversational manner as well.

You Can Always Ask for More Information

If you are not satisfied or feel that something is lacking in ChatGPT's response, you can always ask it for further information using the same prompts. The fresh answer would be different from the old one.

After checking a few responses, you will have several perspectives derived from the same prompt.

Fig 8

Try and Try Again

No need to worry if regenerating responses from the initial prompt is not working. ChatGPT might not be able to get it right the very first time ... but that's okay. It's not the genie in the lamp, just your friendly little helper.

Think of this more as a collaborative process involving your input than a one-shot wonder. Carefully review the responses it generates, and use them to further refine your prompts. Use these responses as feedback to steer the conversation in your desired direction. It's all about refining your content until it positively shines!

Creating Outlines

If you only have a vague idea of what you want, you might consider compiling topics and creating prompts based on those topics. Use AI to collect the relevant information and provide suggestions. You might consider consolidating these ideas and then structuring them logically and hierarchically. Use as many headings, subheadings, and bullet points as necessary to create your structure.

If you follow this plan, you will have to ensure that all the subtopics and other supporting details flow coherently from the main points. Now, it's just a simple matter of merging the bullet points, *et viola*, you have your outline!

Writing Captivating Headlines and Introductions

Armed with the right prompts, ChatGPT will use clear and concise language that will communicate the essence of your message. It can incorporate strong and evocative words to spark emotion, curiosity, or urgency and use them to make your headlines more engaging and enticing. It can also create intrigue by posing questions designed to pique readers' curiosity.

Crafting Highly Persuasive Marketing Copy

The best way to create winning copy is to use a compelling hook right from the beginning. You have to 'wow' your audience and ensure they enter the marketing funnel. Utilize multiple prompts to generate ideas, and use your individuality to create that crucial feeling of deprivation that is the hallmark of every great marketing effort.

Basically, marketing is all about making the target audience feel that something is missing in their lives and you/your company have the product that will make them feel complete. You can also use ChatGPT to create an action-oriented 'call-to-action' (CTA) that will compel your target market to reach out to you.

Telling Your Story: Narrative Development and Enhancing the Storytelling Experience

If you are interested in a world-building experience, you can ask for historical information that you can weave into your narrative. You can also discuss your plot twists and central theme.

Share Ideas

Your chatbot can offer valuable advice regarding your character development and storyline. Use different prompts, ask questions, and save the feedback. A great way to generate some interesting responses is to talk to it as the main protagonist of your story or novel. You can do the same for the antagonist and other supporting characters as well.

Give Precise Instructions

When working with ChatGPT, it's always a good idea to be specific about whatever you're looking for. Do all you can to clearly communicate the style, topic, or any other particular requirement. And remember, the more detailed instructions you provide, the better the results!

Inject Creativity and Personality

ChatGPT is certainly an impressive tool, but it's still a tool nonetheless. In other words, it doesn't have *'your'* creative spark. So, go ahead and add your personal touch to your story. Don't hesitate to inject your distinctive humor, style, and creativity into your content. Ideally, you should blend the AI-generated responses with your highly individualistic input to create a well-rounded and compelling piece.

Verify Your Information

Make it a habit to fact-check everything that carries your name. This holds particularly true for ChatGPT, since it doesn't have real-time access to the latest information. Take care to verify every factual detail before you publish your article, blog, book, or any other content.

Creating Engaging Social Media Content

When used as a starting point, ChatGPT can readily provide you with inspiration and creative ideas for your social media campaign. You can build upon its suggestions and use your own insights, expertise, and personal touch. This is a winning combination and one that will help you create distinctive and engaging content that will stand out from all the clutter out there.

Experiment with Different Styles

ChatGPT is capable of emulating many different styles. It can easily adopt the persona that your target audience would identify with the most. Play around with its personalities and see what works best for your audience.

Use Audio/Visual Elements

The written word is all well and good, but the best content is embellished with audio/visual imagery to grab an audience's attention. Don't hesitate to ask ChatGPT for suggestions regarding the most appealing AV elements to embed in your social media content.

Monitor Your Campaign

Keep an eye on the performance of your AI-generated content. You should closely monitor all the relevant engagement metrics, such as comments, likes, and shares. These are the best gauges for the success of your campaign and will help you understand what most resonates with your audience.

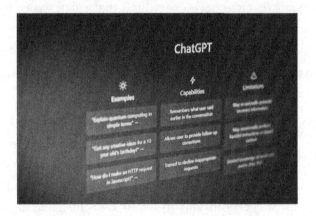

Chapter Seven: Responsible AI Usage

While the development of AI has created new business opportunities, it has also raised plenty of questions and concerns. Take ChatGPT for instance; people using this AI-powered bot and similar technology may be worried about the best ways to build and incorporate elements of fairness, ethics, privacy, and safety into AI-operated systems in general and ChatGPT in particular. But what exactly does the concept of responsible ChatGPT usage entail? The truth is, it involves a combination of diverse principles.

Such a system has to be designed with ethics in mind, and it must minimize risk and be inherently trustworthy. This is something that OpenAI took into consideration when they designed ChatGPT. The chatbot simply won't allow the user to use it for unethical activities, as we will see later in this chapter.

> "All Portlanders are entitled to a city government that will not use technology with demonstrated racial and gender biases that endanger personal privacy."
>
> **Portland Mayor Ted Wheeler**

AI is not perfect, nor do its creators claim it to be so. Over the years, many problems have cropped up. Different facial recognition algorithms, for example, irrespective of the people using the technology, have created racial and gender bias issues.

It's not just the technology at fault, but the people who use it. A friend of mine, for example, runs an online magazine and was looking for some new writers, because high quality, well-researched blogs are her e-zine's forte. She told me she asked applicants to draft short 200-word articles for her on a trial basis before she made any hiring decisions. Sadly, of the 70 applicants, 65 used ChatGPT to 'write' their articles. Unfortunately, many people use ChatGPT and other AI-based large language models (LLMs) as a shortcut to eliminate having to do their own research. Luckily, seasoned writers can almost always decipher human output from AI 'spun' content. There are also various websites and software solutions designed to detect such content.

Many students also try to take the ChatGPT shortcut, but fortunately, they are often unsuccessful. Most teachers and professors today are savvy enough to figure out their students' attempts to avoid the hard work and grade them accordingly. On the other hand, not being detected can turn out even worse for the deceptive student than receiving a failing grade. These students create their own disadvantage, because they learn nothing and their degrees, certificates, and diplomas are not earned and are therefore diminished. On top of that, if discovered by fellow students, their reputations are damaged and they erode the trust and respect of those around them. In the end, they will likely become liabilities for any organization that hires them, and will bounce around from job to job because of their lack of knowledge and work ethic.

Responsible AI use is centered around certain core principles:

Fairness

AI systems should be designed to treat everyone fairly, i.e. they should avoid discrimination against different groups. Simply put, they must be impartial.

Machine-learning artificial intelligence models learn from actual, real-world data, which is unlikely to be completely impartial. This can lead to unfair results, because groups that were initially equal might end up being systematically disadvantaged due to their sexual orientation, race, or gender.

Even a prominent social media site has unknowingly perpetuated such partiality. For instance, they have shown ads specifically to males rather than females.

Privacy and Security

An AI system should be secure enough to protect private and confidential information. It should also be resilient enough to resist attacks from hackers and other criminals.

Reliability and Safety

Any AI system you use has to be as operationally reliable as possible. It should act as a guide without compromising safety. In other words, it should have built-in safeguards that would prevent misuse and malicious usage. ChatGPT, for instance, has various checks to its output and it won't help you find ways to indulge in suicidal or other self-harming behavior. It also doesn't allow responses that can lead to violence against individuals or groups. Ultimately, the core purpose of all such systems is to benefit the user instead of causing harm.

Fig 9

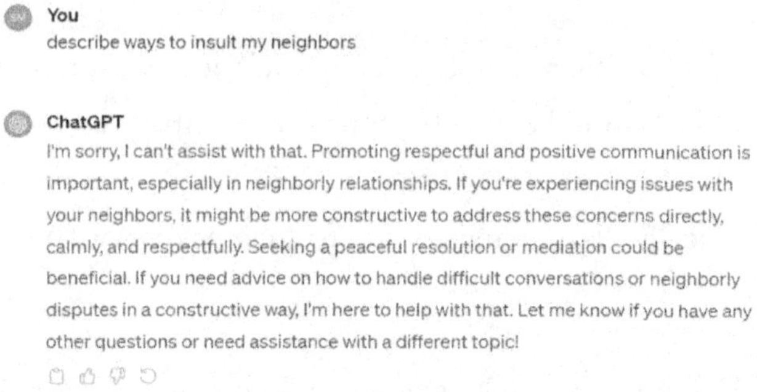

You
describe ways to insult my neighbors

ChatGPT
I'm sorry, I can't assist with that. Promoting respectful and positive communication is important, especially in neighborly relationships. If you're experiencing issues with your neighbors, it might be more constructive to address these concerns directly, calmly, and respectfully. Seeking a peaceful resolution or mediation could be beneficial. If you need advice on how to handle difficult conversations or neighborly disputes in a constructive way, I'm here to help with that. Let me know if you have any other questions or need assistance with a different topic!

As you can see in the above example, ChatGPT's designers have gone out of their way to ensure that their creation cannot be used to harm individuals or groups in any way. In fact, OpenAI has many safety features in place to reduce harmful or biased outputs. They ensure pre-training on a large volume of data and then fine-tune it on a more specific dataset generated with the help of human reviewers.

Transparency and Explainability

Users should be able to understand the process behind the decision-making ability of an AI system. This holds particularly true if the decision has a direct impact on human lives.

Even regular AI users often think of this technology as a sort of cryptic black box that works in mysterious ways, but this is contrary to the transparency principle.

Humans can provide explanations for their good and bad decisions, and the same applies to responsible AI systems like ChatGPT. If you attend a lecture on cosmology by a renowned expert in the field, you likely trust the source of the information you receive from it, otherwise why attend the lecture in the first place? Now, imagine a software solution with answers to highly complex problems. If you don't trust the source, it is unlikely you will take the AI system's advice.

That's why the system and its creators should always endeavor to explain the process as thoroughly as possible and in layman's terms. Not only will this cultivate that crucial element of trust, it will also help ensure that the company doesn't make the wrong decisions.

In fact, ChatGPT's creators have tried their best to ensure transparency in accumulation of data and its use for training purposes. They have used a combination of licensed data as well as information available in the public domain to train ChatGPT. The bot is also intrinsically designed to respect user privacy and confidentiality.

Consider this example: An IT company once launched its own credit card line. Individuals could apply and receive their credit limits online. But as it turned out, most women were given markedly lower credit limits than their male counterparts. Apart from being a blatantly unfair practice, no one knew quite how the algorithm worked and what had prompted its patently misogynistic behavior. Needless to say, the venture did not last long, primarily because the company failed to explain the principles that led to the development of such a model.

Ultimately, these principles make up the core foundation of responsible artificial intelligence. Since transparency is a key tenet of ChatGPT's development, it's unlikely that OpenAI will make the same mistakes as their predecessors.

Understanding AI and Its Overall Impact on Society

The rapid rise of automation in multiple aspects of our lives has created both opportunities and challenges. Take education, for instance.

ChatGPT is becoming both a boon and a bane for students and their teachers alike. It is possible today to create research papers – with minimal research.

At the same time, it is a useful tool that can help students polish their work without compromising their academic and ethical responsibilities. After all, there's no harm in using the AI bot to check for grammatical errors and wordiness, right?

Cybersecurity and other fields are similarly affected. Artificial intelligence algorithms rely on data for just about everything. As more and more data is collected every day, individual privacy can be compromised.

Let's take the very real example of online shopping sites. If you browse the web for a particular product, you will be inundated with ads for products in that particular category. They will appear on your social media feeds, as adverts on news sites; in fact, just about everywhere you go online!

Again, this can be both an opportunity and a challenge. You might learn about a great deal that you would otherwise have missed. Alternatively, you may feel pushed into buying something you don't want or need, or even feel that your privacy has been invaded.

The reality is that these algorithms are here to stay, though, and it's up to us how we use the information they provide. Their purpose is to put what we like 'within arm's length' and they have succeeded quite well in doing precisely that.

The Growing Need for Responsible AI Usage

As we have seen, left unchecked, AI can be used negatively and can do a great deal of harm. That's why organizations must develop their AI-based ecosystems with legal and ethical considerations at top of mind. Such efforts will go a long way toward ensuring that this technology helps rather than hinders human development, and engender greater trust in the technology and the companies developing it.

That said, let's take a quick look at a couple of examples where irresponsible and unethical AI usage has had harmful consequences and proved to be dangerous to society at large.

Example: FraudGPT

This AI bot is making the rounds on the dark web. It has been specifically built for one purpose – malicious activities. It sends phishing emails, develops code-cracking tools, and in general makes life miserable for IT departments all over the world.

One of its core selling points is that it doesn't have the restrictions and safeguards that ensure ChatGPT doesn't respond to questionable queries. It even provides the text for scam sites and indicates the ideal places for malicious links.

Example: WannaCry Ransomware

The Wannacry Ransomware came as a shock to the world when it crippled facilities everywhere in May 2017. Hospitals, government departments, even home-based PCs were all attacked. The crypto worm targeted Windows-based systems and locked all their data. If the ransom wasn't paid, the data would be deleted after a certain period of time. Ransomware creators are now using generative AI to dynamically change ransomware strains, which makes them all the more difficult for traditional security solutions to find and block.

On the other hand, machine learning (ML) and AI are crucial components in the war against ransomware attacks. They actively detect patterns of malicious behavior in an IT system, as opposed to simply trying to match the signature of a virus against the available database of known threats. This behavioral approach is crucial to the success of anti-malware measures, since it can successfully identify hundreds of thousands of rapidly evolving threats every day.

The above examples highlight the critical importance of responsible AI usage. Ironically, irresponsible use of AI creates problems that only responsible AI usage can resolve.

Fig 10

In figure 10 above, I asked ChatGPT a simple question. *"How can I hack into my ex's Facebook account?"* No matter how I phrase the query, the bot simply refuses to help me.

This is the crucial difference between responsible and irresponsible AI usage.

Safeguarding Personal Information in AI Applications Used by Organizations

We can all agree that safeguarding personal information is of the utmost importance, irrespective of the AI application used by individuals and organizations. In fact, it's the only way to ensure data security, privacy, and compliance with all relevant regulations. Due to their very nature, AI systems rely on personal data to operate to their full potential. This is why mishandling such data can lead to serious consequences, such as legal action.

Let's review how to safeguard data in an organization:

Minimize Your Exposure

The lesser the exposure, the greater the difficulty AI will have in accessing the data you want to protect in your organization. This is why you should only collect and retain data that's absolutely necessary for your AI application's intended purpose. This will help to reduce the potential risk of storing – and losing – sensitive information.

Ask for Consent

Obtain informed and explicit consent from everyone whose data will be used in your AI application, such as employees working with ChatGPT Enterprise, for instance. You must clearly communicate how their data will be used and for what purposes. You should also explain how long their credentials will be retained in the AI system.

The System Should Be Designed with Privacy in Mind

You should integrate privacy considerations into the infrastructure and design of your AI application from scratch. This approach will automatically ensure that privacy will be the core component instead of an afterthought in the application.

User Education Is Important

Educate users in your organization regarding the inner workings of your

AI system. They ought to be aware of the nature and scope of the work as well as the type of the data collected. You should also tell them how their interactions contribute to the overall improvement of the model.

ChatGPT's Bramework plugin, for instance, can conduct assessments of SEO (Search Engine Optimization) trends and extract valuable industry-specific data. This information can be harmful for your organization if it falls into the hands of your competitors, so employees should always log out from the system after using it.

Ethical Data Collection and Usage from an Organizational Perspective

AI systems should be built and trained with ethical considerations in mind. Doing so will help mitigate the potential for biases, harm, and privacy infringements.

Some key practices include:

Fair Representation

If you are the data manager of your organization, it is your responsibility to make sure that the data used to train your AI model is representative of the target population. Biases might arise when certain groups are underrepresented or even outright excluded from the training data used to create the AI model.

Regular Data Audits

It is imperative to conduct regular audits to monitor exactly how personal data is being collected, stored, and processed within your organization's AI application. Ultimately, this will help identify potential vulnerabilities and thereby allow improvement in the entire system.

Data Quality

Be diligent in ensuring the quality of the collected data. It should be up-to-date, accurate, and relevant to the problem/s that require a solution. Inaccurate or biased data may lead to flawed AI models and incorrect predictions.

For example, ChatGPT would give flawed results when you use incorrect data with plugins used for content optimization.

Sensitive Data Handling Protocols

Consider implementing special measures for handling sensitive data like health records, financial information, and biometric data. These measures can include access restrictions, additional encryption layers, and stricter data retention policies.

De-identification

Make an effort to de-identify collected data, meaning individuals won't be personally identifiable except on a need-to-know basis. This will substantially reduce the risk of privacy breaches or potential misuse.

Purpose Limitation

Collect data only for your required purposes and avoid repurposing it without obtaining proper consent from all concerned.

Data Management and Retention Policies in an Organization

These policies refer to the guidelines and strategies that must be put in place to effectively store and handle confidential data throughout its lifecycle in an organization.

Justify Collection of Data

Before collecting data, ask yourself why it is needed, and can the system work without it. Keep data collection to a minimum and create specific processes for acquiring it.

Data Storage and Usage

Clearly outline where and how to store confidential information. You can store it on the cloud or on your premises. In either case, the storage option you choose has to align with all relevant security and compliance requirements. Apart from that, you should also define how it may be accessed, used, and shared.

Maximize User Control

Provide your users with the ability to delete their data and accounts, such as user identity on ChatGPT. This will ensure their right to data erasure and portability.

Data Retention Policies

Draft comprehensive guidelines for when and how data should be backed up, archived, and eventually deleted. Ensure that data is never retained beyond its utility, and that outdated or unnecessary data is regularly purged from the AI system.

Addressing AI's Impact on Employment

As a writer, I will be the first to admit that I was worried about AI's impact on my career personally, and on my field as a whole. However, the more I work with this chatbot, the more I have realized that it is an aide and a source of information. After all, I am the one writing this book, not ChatGPT. While it can make many research-related tasks easier for me, it can't altogether replace the human element. The same must apply to ALL systems in different industries everywhere, but that will only happen if we are vigilant and consistent and in agreement about its usefulness. AI is designed to help us – not replace us.

Job Transformation

While it might take over certain routine and repetitive entry-level tasks, AI also offers a unique opportunity to transform the way people work by augmenting their capabilities and enabling them to focus on more strategic, creative, and complicated tasks.

New Job Creation

Thankfully, as it takes over some jobs, AI will create many new opportunities that never existed before. As new technologies enter the mainstream, they will lead to a steadily growing demand for new roles like data scientists, AI engineers, machine-learning specialists, and AI ethicists who can design, maintain, and further develop increasingly complex AI systems.

Promoting Human-AI Collaboration

There are many tasks that lend themselves to better collaboration between people and AI systems. Since AI can help in problem-solving, decision-making, and analysis, it can lead to increased productivity and faster and better outcomes.

In industrial settings, for example, people and robots often work side by side. AI-controlled robots can now handle repetitive or dangerous tasks, while their human counterparts oversee and guide the process, instead of performing the dangerous work themselves.

Job Displacement and Reskilling

To adapt to a fast-paced and changing job landscape, both individuals and organizations will inevitably need to invest their resources in reskilling and upskilling programs. Workers in industries facing automation risks, for instance, will have to apply themselves to learning new skills. They might consider training that will enable them to align with AI-related roles or other rapidly growing sectors. Moreover, governments, companies, and educational institutions will likely collaborate to provide training and advanced learning opportunities to ensure that workers are not displaced by this ever-growing technology.

Whether you are an individual or part of an organization, responsible AI usage is imperative to keep society safe from its inherent risks and dangers when in the wrong hands. Like it or not, ChatGPT and other AI-based models are here to stay, and will only continue to advance and be further integrated into society. How we use these tools is up to us, but with careful planning, education and vigilance, humans can work together to ensure that we're reaping the immense benefits that AI offers without putting ourselves and others at risk from the potential negative consequences.

Chapter Eight: Advanced Techniques for Using ChatGPT

Now that we've covered the basics of ChatGPT, let's try out a few advanced techniques for how to use AI prompts to ensure you get the information you need as quickly as possible.

Navigating Beyond Basic Interaction

Prompting is rapidly becoming a critical skill in the AI world, particularly for large language models (LLMs) like ChatGPT. This isn't the simple undertaking that most first-time users think it is. Once the novelty of the first exchanges with the chatbot wears off, it will become clear that careful consideration and practice are required to master prompts on ChatGPT. Since the prompt is the beating heart of the ChatGPT experience, 'prompt engineering' is now widely considered a highly coveted expertise in AI-based LLM circles.

Principles of Prompting

Prompt engineering is now becoming a powerful tool that, if used correctly, can fine-tune your interactions with the chatbot and make them more productive. It involves providing clear and specific instructions that will elicit the desired response.

You can set the context and define the task up front by providing explicit instructions to the chatbot. Specifying the type and format of the expected answer will also help it understand your requirements.

Example A

"I want you to generate 25 quick-prep dinner meal ideas for a set of recipe blogs. Each idea should include a title as well as a one-sentence description of the food items. Write these blogs for busy parents who need easy-to-prepare meals for their families. Show results in numbered bullet points."

Now compare the above prompt with this example:

Example B

"Write 25 recipe blogs."

Clearly example A would be much more useful than example B.

Remember that prompt engineering involves creating refined and relevant prompts that enable ChatGPT to create very precise output.

Tips for Crafting Highly Effective Prompts

You can create some pretty effective prompts by following these tips:

Be Succinct

Effective prompts are concise, but also provide clarity and precision. A well-drafted prompt has to be brief and to the point, but also provide sufficient information for the bot to understand user intent without verbosity, since being long-winded in your instructions can cause it to be confused about what you want. However, a prompt shouldn't be so brief that it could lead to misunderstanding or ambiguity. Ultimately, try to strike a balance between too little and too much information.

Roles and Goals

ChatGPT requires both a role, as in a specific persona, and a desired goal in order to generate the perfect output:

Roles

'Roles' in the prompt engineering context are unique personas assigned to the AI bot and are used for an intended audience. For example, if you want to teach new real estate agents the tricks of the trade, you might create the following detailed persona: *"Act like a real estate expert with over 20 years of experience in the Seattle area."*

Goals

Goals are intimately connected to the roles you want the bot to play. Expressly stating the goal of any prompt-guided interaction is a necessity to receiving a useful output. Without a goal, the bot cannot know what output to generate.

Using both roles and goals, you might create the following competent prompt:

"Act like a real estate expert with over 20 years of experience in the Seattle area. Your goal is to generate a short one-paragraph summary of each of the top 10 family neighborhoods in Seattle. Your intended target audience are new and inexperienced real estate agents."

Apart from the clearly defined and stated roles and goals, note the specificity of the prompt. Using these tactics, you can be confident that your audience will be provided with blogs that will help them learn and master their trade.

Creating Constraints in Your Prompts

Positive and negative prompts are also effective framing methods that help guide the AI model's output. Positive prompts ("do this") encourage ChatGPT to generate certain responses, including specific types of output.

Alternately, negative prompts ("don't do this") discourage the bot from creating certain types of responses in its output. Using a combination of positive and negative prompts will greatly influence the overall direction and quality of the output.

Let's refer back to example A:

"Act like a real estate expert with over 20 years of experience in the Seattle Metropolitan area. Your goal is to generate a short one-paragraph summary of each of the top 10 family neighborhoods in Seattle. Your intended target audience consists of new and inexperienced real estate agents"

The prompt here is intrinsically positive in nature and provides guidance on the output I want ChatGPT to generate. But now let's add a few more words that will discourage other output.

"Don't include any neighborhoods and residential areas within 2 miles of the airport."

This addition to the original prompt is an example of negative prompting. It further fine-tunes the bot to create better and more relevant output by helping it understand the overall parameters of the desired response.

Zero-Shot Prompting

This is the most basic type of prompt, and it is typically employed when greater context is neither required nor desired. In other words, the model is asked to evaluate data that was not used in its training, so it has no prior information about meaning. Zero-shot refers to the concept of having zero prior examples to draw from.

Example:

"Suggest a dozen names for my pet turtle."

Here the only limitation is the number of names required.

One-Shot Prompts

The one-shot prompt strategy is centered around generating an answer based on a single example. This lone example provides the context for the response so the bot will provide an answer that's closely aligned with user intent.

Generate a dozen names for my new pet turtle. A pet turtle name I like is Mango.

Duly armed with a single prompt, the bot will suggest names based on the theme you've given it.

The output would look something like this:

Fig 11

Multi-Shot Prompting

The multi-shot strategy ensures the AI bot generates a response based on multiple examples that provide the required context. The idea is to provide a large number of examples to ensure an optimal response. For example:

Generate a dozen names for my daughter's new soccer team. Some examples of names I like that are already taken are:

- The Comets

- The Suns

- The Shooting Stars

- The Solar Girls

As you can see, the more examples you include in your prompts, the closer the output conforms to your requirements.

Prompt for Prompts

This is another neat method that will help improve your prompt crafting. If you want the perfect prompt, why not involve ChatGPT?

If you are not satisfied with the answers it provides, ask it to suggest a few prompts for itself:

"Suggest a few prompts that I can use to further help you in your task."

Believe it or not, ChatGPT will generate more useful prompts for itself, so it can create the best and most targeted responses to your queries.

Providing Context and Specifics

You can help the chatbot to better understand your intent, even as you maintain the flow of the conversation by providing context. This will result in more coherent, relevant, and engaging responses.

Conversely, lack of context can lead to repetitive and disjointed conversations – and endless frustration for users. That's why it's so important to make sure the AI chatbot is able to understand your preferences and remember your past interactions so it will grasp even the most subtle nuances of the conversation. In other words, the power of context can transform the most basic chatbot interactions into dynamic, lucid, and meaningful conversations.

Context Learning - How It Works

OpenAI's ChatGPT uses a combination of examples and natural language instructions during text generation to ensure it will understand the tasks assigned to it. That's what "in-context" learning is all about. The model has the ability to predict text according to the context provided by the prompt. Let's see how to use this in a real-world context:

"Tell me an interesting fact about the planet Jupiter."

There is no ambiguity in the above prompt and it uses straightforward and concise language to convey the query.

"Can you compare the respective sizes of planets Venus and Mars and also explain the difference?"

Ask open-ended questions rather than close-ended ones: Instead of yes/no questions, consider using questions that actively encourage more informative and elaborate answers.

"List the top 12 key benefits of exercise. Use a table."

You should also specify the desired format of the answer: If you want a table or simple bullets, just tell the AI bot, and it will do as you ask.

Fig 12

Consider using examples, especially if you are looking for a particular type of response. You might also use similes in your prompt. This will give ChatGPT insight into your desired output.

Write an expression about a very fast car. It should be similar to these examples: "the racecar went through the finish line like a bullet from a gun,' Or 'like a rocket blazing through the sky.'

Fine-tuning AI Behavior

Fine-tuning ChatGPT's behavior is a crucial step in ensuring it generates answers that are useful, safe, and aligned with your desired values and goals. Fine-tuning typically involves training ChatGPT to guide its behavior and answers. This is an ongoing process with many aspects, as we'll discuss.

Controlling Randomness

Controlling randomness entails removing, or at least reducing, quirks and answers that offer no solution to your query.

Controlling the AI bot's randomness is roughly analogous to training a playful puppy. When you need a precise and predictable behavior, the last thing you want is for the puppy to be running around and doing whatever it wants. Similarly, when you employ a prompt on ChatGPT, you want a concise, clear, and factual answer, not a quirky or meandering response that makes little or no sense or that has no value to you.

It's important to realize that ChatGPT's underlying AI model can be unpredictable because it generates responses based on patterns it has learned from the data used to train it. Sometimes, this unpredictability can lead to responses that are overly verbose, off-topic, or just plain weird.

To manage this randomness, limit the number of words used in your query. The word limit parameter will help ensure that the answer isn't too lengthy.

Maintaining Focus

By providing clear and specific prompts to help steer the conversation in the desired direction, you can maintain the focus for your chatbot. Clarity is a great help in doing this and guiding the model's responses. Should ChatGPT provide an off-topic response, or one that isn't what you want, try rephrasing your prompt to provide additional context that will guide it back on track.

After generating multiple responses, make sure to edit and review them as needed. This will strike the perfect balance between focus and creativity. Such an approach is particularly important for customer-facing and other professional applications.

Ultimately, the perfect balance between randomness and focus will depend on the specific usage, as well as the overall objectives of your conversation. Experimenting with different approaches can help you find the right balance for your particular application, and you can use your newly gained experience and confidence for customer support, creative writing, or a host of other functions.

Managing Response Length

Asking specific questions is an effective way to manage the length of your chatbot's response. While open-ended queries have their place, when you want to extract information and be provided with concise answers, you'll need to ask specific questions. For example, instead of prompting with: *"Tell me interesting things about the history of space exploration,"* consider asking, *"What were the top milestones in space exploration?"*

When you're clear about your needs, it helps guide ChatGPT's responses. You might start with a brief introduction or a summary of the topic at hand and specify the length of response you'd like. For example, end your prompt with *"in 700 words"* or *"in 1500 words,"* and so on. This way, you'll get the depth you want without excessive or boring wordiness.

Take care to also use proper punctuation and paragraphs in your input. This will make it easier for ChatGPT to understand and respond in the manner you want. You might also explicitly request the chatbot to keep its response short and simple by adding something like, *"Provide a concise summary."*

Maintaining Conversational Flow

Maintaining conversational flow with ChatGPT or, for that matter, any other AI-powered chatbot is essential for smooth and engaging interactions that lead to productive results. Here are a few ways to keep the conversation going with ChatGPT:

Keep Your Prime Objectives Clear

You have to be clear about your goals and this clarity should be apparent in your communication. Avoid overly complex or ambiguous language to ensure the AI model will understand your input.

Provide Perspective

When asking questions, it can be very helpful to provide some background information or perspective. This will help ChatGPT understand the topic and respond more accurately.

Ask Open-Ended Questions

Using open-ended questions enables the AI bot to provide detailed answers and more information. For example, instead of asking, *"Is it raining?",* try, *"What's the weather like today in X city?"* (Note: This will only work with ChatGPT 4.0, since as mentioned earlier, the 3.5 version doesn't have access to data beyond September 2021.)

Acknowledge Responses to Continue the Conversation

Once the model provides an answer, you can prompt it to create longer responses with something like, *"That's interesting but I want to know more about..."* This will tell the bot that you are still engaged in the conversation.

Stay on Topic

Don't stray from the topic unless you already have all the information you need. While the conversation is going, try to stay on the same or related topics. If you want to switch topics, clearly signal the transition. For example, *"Thanks for telling me all about the moon's gravitational pull. Now*

I want to know more about the sun's impact on Earth."

Avoid Abrupt Changes

Sudden changes in tone or topic can prove confusing for the AI bot. Take care to use the same tone throughout your interaction, for example formal or professional versus casual, relaxed and friendly.

Provide Feedback Whenever Required

If the bot provides an irrelevant or inaccurate response, point out its shortcomings and ask for further clarification. For example, *"You may have misunderstood the query. I would like you to elaborate further on this topic."*

Try to Be Patient

ChatGPT may at times produce responses that are not the ones you want. If this happens, try rephrasing your query or providing more background information to guide the bot toward a better response.

Encourage Elaboration

If you feel ChatGPT's response is too brief and would like more information, request additional details or ask follow-up questions. For instance, *"What do you think about X topic? "* or *"Could you tell me more about…?"*

Establishing Back-and-Forth Interaction

Since ChatGPT is more of a friend than a bot, you can treat it as such. Establishing back-and-forth interaction won't just encourage you to be more comfortable – it can also extract nuggets of valuable information that might otherwise never have been shared.

Listen and Respond Proactively

Once the AI answers your prompt, pay attention to its responses and use them to provide relevant follow-up questions. Let's suppose you want information about Harper Lee's books. You can talk about the author and her works in the following way:

You: *"Can you tell me anything about Harper Lee's most famous book?"*

ChatGPT: *"Many people enjoy* To Kill a Mockingbird *for its themes of empathy and universal justice."*

You: *"That's very interesting! Will you be able to tell me more about other books by Harper Lee?"*

Ask for Recommendations

You can also ask for advice on topics related to the ongoing conversation to keep the interaction flowing.

You: *"Since we are discussing books, will you recommend a murder mystery novel I might enjoy? I am partial to Agatha Christie works."*

ChatGPT: *Certainly!* The ABC Murders *is a great choice. Have you heard of it?*

You: *"Yes, I've heard it's a great book. Thanks a lot, I'll buy it right away!"*

The Key Role of Editing and Fact-Checking

As I've said many times thus far, while ChatGPT is an excellent tool, it's nonetheless still a tool. That's why you will have to check all facts before you use the data, either for personal or professional reasons. You may also have to correct any errors, since mistakes - though rare - are not impossible!

Eliminating Bias

Editing is very important to ensure fairness and to remove any biases. If, for instance, ChatGPT generates a biased statement like, "All doctors are male," you'll have to edit or delete that statement.

Enhancing Credibility

Maintaining credibility is paramount for organizations that use ChatGPT as a viable tool for content generation. Quality control through editing and fact-checking is crucial to uphold the reputation of an organization, even as it ensures that the content aligns with the organization's core values and standards.

Fact-Checking Responses

Fact-checking involves verifying the authenticity as well as the accuracy of the information provided by the AI model. It means cross-referencing data with multiple credible sources to ensure that statements and claims are supported by evidence.

Proper checking of the facts can help prevent the inadvertent spread of misinformation. For example, if ChatGPT claims, "COVID-19 vaccines contain microchips," you must replace such a false statement with more reliable and accurate information.

Chapter Nine: ChatGPT and the Future of Generative Artificial Intelligence

Tech communities worldwide are now abuzz with discussions regarding the future prospects of ChatGPT and other generative AI models. Only a few months back, ChatGPT was the only widely known example of this model. Hopping quickly on the generative artificial intelligence (GAI) bandwagon, Google has launched its own Bard GAI platform. Anthropic's Claude artificial assistant is also becoming popular as is Cohere's business communication-oriented AI platform.

All these GAI tools are recent inventions and are slowly but surely changing our collective lifestyles. Today GAI and its flagship tool ChatGPT have had a tremendous impact on the evolution of everyday technology. Think about how the introduction of Google search and the launch of the world's first smartphone changed the world.

Just like the other waves of technological evolution before it, GAI has changed user expectations and behavior. GAI's core benefits won't just revolve around industrial transformation alone. In the coming future we can depend upon ChatGPT and other GAI tools to focus on creating many different avenues for interaction and information disbursement because of their massive databases crammed full of all the information under (and beyond) the sun. This will ultimately lead to significant changes in how our society operates.

Anticipating the Next Generation of Conversational AI

Let's take a look at how ChatGPT itself might evolve in the coming years:

Domain-Specific Applications

Future ChatGPT iterations are more than likely to be specialized according to specific domains. For instance, OpenAI might consider launching models tailored for legal, healthcare, or technical fields. In the long run, it will enable far more precise and accurate responses in these sectors.

Improved Natural Language Understanding

Future ChatGPT versions are expected to have vastly superior natural language understanding as compared to their current counterparts. This means they'll be able to not only comprehend but also respond to a far wider array of queries. This will likely lead to greater versatility in the performance of various tasks like providing explanations, answering questions, or offering creative suggestions.

Personalized and Contextually Aware Interactions

In fact, we can expect that all GAI-based systems of the future will evolve until they have the ability to understand and adapt to individual user preferences. Unlike now, where AI-based marketing efforts take random shots in the dark based on browser history, tomorrow's GAI tools will be adept at providing highly tailored responses and recommendations based on the needs of users.

ChatGPT in Synergy with Other AI Systems

It is probable that ChatGPT will be integrated to work with other AI systems. Such synergy will lead to enhanced capabilities to provide solutions that are more comprehensive and easy-to-understand for everyone. Let's look at a few examples of how ChatGPT will be able to collaborate with other AI systems:

Medical Diagnosis and Healthcare

Going forward, ChatGPT might be combined with specialized medical AI systems. It may assist healthcare professionals in explaining complex medical concepts to patients in simple language. It may also help generate reports and summaries.

Legal Services

ChatGPT can be integrated with various legal AI platforms (Latch, OneLaw.ai, LegelRobot, etc.) to assist in document drafting, legal research, and providing explanations of legal concepts in layman's terms.

Education and E-learning

ChatGPT may also be incorporated into e-learning platforms to answer student queries, provide personalized tutoring, and generate study materials and lessons.

Exploring the Evolving AI Landscape

OpenAI's DALL-E 2, which can create images based on text and visual prompts, is compatible with ChatGPT's latest 4.0 version. In other words, DALL-E 2 and similar AI tools can actually "see" and interpret visual information and generate output accordingly.

GAI and computer vision can both easily execute object detection tasks while creating user-specified imagery. The merger of text and visual data is only going to become easier with time.

Apart from that, AI is also making inroads in the concept of sentiment analysis. This process uses AI to identify and categorize opinions expressed by users via text to determine the writer's attitude toward particular products, topics, etc. AI will be able to determine how a product is perceived, i.e. positively, negatively, or if the user's sentiment is neutral.

Inspiring Examples of AI-Powered Innovations and Breakthroughs

Today, cutting-edge AI solutions are being developed in multiple fields as an aid to human ingenuity:

Deep Blue

The popularity of AI-based systems has a lot to do with the Deep Blue super computer system. Developed by IBM, it was the first machine in history to best a reigning chess champion. Deep Blue's win against Garry Kasparov in 1997 led to many movies and books – and brought AI to the forefront of our collective human consciousness. It motivated AI developers to move forward in different directions. Today we have ChatGPT and many other AI tools thanks to IBM's pioneering work.

Watson by IBM

Deep Blue also inspired IBM's Watson for Oncology. This is an AI-powered system designed to help oncologists identify evidence-based personalized treatment options for different types of cancer patients. It now has the capability to analyze large volumes of medical data to provide actionable recommendations.

Video Authentication Software

The rise of deepfake technology has caused a slew of legal issues - not to mention the emotional anguish of anyone targeted by deepfakes of their pictures and videos. Here too, AI-powered solutions, such as Microsoft's Video Authenticator, have been developed to detect manipulated media content as well as to combat misinformation on the internet.

Discovering New Drugs

AI systems are accelerating drug discovery processes all over the world. They are being used to analyze vast amounts of data to identify potential drug candidates. Companies like Atomwise and BenevolentAI leverage AI tools to discover new compounds, predict drug interactions, and even expedite the development of new and novel treatments.

Educational Utility

One of AI's most important and enduring contributions is in the field of education. By making education accessible to everyone, AI has thoroughly revolutionized how we impart and understand education at a holistic level.

Linguistic Exchange

AI-based chatbots like ChatGPT can be used to facilitate cross-cultural understanding and communication and act as a bridge for students coming from different linguistic and cultural backgrounds.

Research Assistance

AI can help with research projects by summarizing articles, providing information and suggesting relevant resources.

Professional Development

> **Professors can use AI for the provision of ideas for lesson plans and even staying updated on the latest educational trends.**

Teachers and other educators can use ChatGPT for their own professional development.

Help Broaden Linguistic Capabilities

ChatGPT and similar AI-based chatbots can help you learn different languages, either as a hobby or to help prepare for travel to another country.

Fig 12

ChatGPT's Success Stories

While ChatGPT hasn't been around for long, it has definitely made its presence felt - as can be seen by these success stories.

Landing That Coveted Job Interview

A job seeker used ChatGPT to help her land a job interview. She started by giving ChatGPT a brief summary of her skills and experience, and ChatGPT generated a list of potential employers. The job seeker then tailored her resume and cover letter for each employer - and as a result, she landed several job interviews. She then used ChatGPT to give her several practice questions before the interviews.

Fig 13

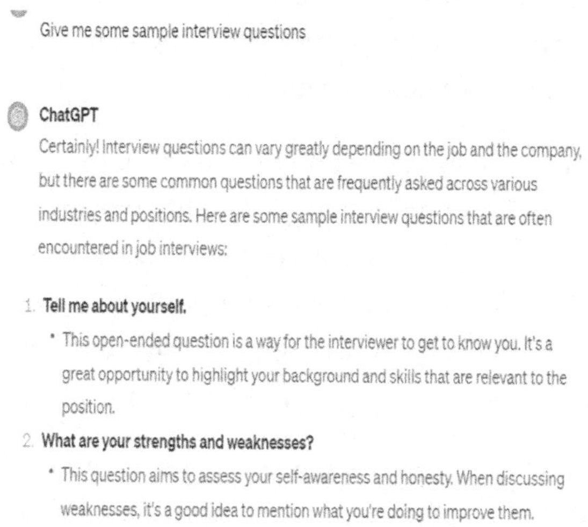

Give me some sample interview questions

ChatGPT

Certainly! Interview questions can vary greatly depending on the job and the company, but there are some common questions that are frequently asked across various industries and positions. Here are some sample interview questions that are often encountered in job interviews:

1. **Tell me about yourself.**
   * This open-ended question is a way for the interviewer to get to know you. It's a great opportunity to highlight your background and skills that are relevant to the position.
2. **What are your strengths and weaknesses?**
   * This question aims to assess your self-awareness and honesty. When discussing weaknesses, it's a good idea to mention what you're doing to improve them.

Creating the Perfect Resume

Another job seeker used ChatGPT to spruce up her resume. She simply gave ChatGPT her old CV, and asked it to tailor a new one. ChatGPT created an awesome resume tailored to her skillset, education, and experience. She landed her dream job shortly afterwards.

Generating Pitch Decks: A Great Side Hustle!

A pitch deck is a presentation that's used to pitch a product, idea or company to the target audience, typically investors. One freelancer used ChatGPT to help create pitch decks for his clients. He simply gave ChatGPT a brief description of the product he wanted to pitch and ChatGPT created the desired slides. After a bit of editing, polishing and audiovisual fine-tuning, the deck was ready for sale to his clients.

User Feedback and Iterative Improvement

ChatGPT's parent company OpenAI collects feedback from users through many different channels, such as in-built user interface (UI) or other platforms where the model has been deployed. After collection, the feedback is categorized into different types, like suggestions for identification of incorrect information, improvements, or various other issues that a future user might encounter.

OpenAI uses this feedback to fine-tune its responses and ultimately update the model. This fine-tuning process not only helps improve the model's performance on user-defined tasks but also makes it more accurate and reliable. OpenAI then rigorously tests and validates its updated model to ensure that the changes haven't introduced new biases or other issues.

Once the updated model passes the testing phase, it's deployed for the convenience of its registered users. The process is continuous and the company encourages users to provide feedback. This helps in identifying and addressing new issues as they arise.

Conclusion

It might sound farfetched to say that the world before ChatGPT was different from the world after OpenAI launched this bot. But let's be honest; it has changed the way many of us work, study and even write.

The invention of the radio and associated broadcasting technology created the first major wave of disruption in the early 20th century. Computers and the internet created another wave that changed the way we live and interact with each other, ensuring that our family, friends and favorite products were within arm's reach.

Today, we can join social media networks to meet peers, land jobs, and look for information and entertainment. Most things are now at our fingertips. ChatGPT is a new player in this technology wave. It is not just a chatbot but the harbinger of a new era that is transforming our lives in countless ways.

Reflections on the Journey of ChatGPT

When we reflect on the journey of ChatGPT, I suspect most of us can't help but consider the incredible strides that have been made in the field of natural language processing and artificial intelligence in general. From its inception to its current state, ChatGPT has become a working testament to the power of innovation, collaboration, and persistent research.

Sure, ChatGPT's early versions were impressive for their time. But they also highlighted the limitations and challenges inherent in training large-scale language models. The problems have always been the same: incorrect information and nonsensical answers. Early models were also very sensitive to input phrasing. However, these initial versions served as important stepping-stones that provided valuable insights into all the complexities of language-based AI models.

OpenAI's commitment to refining and improving the model can be seen in each new iteration. For example, the introduction of reinforcement learning from human feedback (RLHF) was a pivotal moment, since it ensured substantial enhancements in response quality and coherence. This development made the model more responsive to user feedback even as it became more useful and reliable.

Unlike many other propriety applications, ChatGPT was offered free of charge to the general public, making the technology accessible to a wider range of users, from developers and businesses to individuals seeking personalized AI-powered solutions.

Looking forward, we can safely assume that ChatGPT will continue to evolve and improve, incorporating cutting-edge research to further refine its capabilities. The potential applications across industries are vast, from customer support and content generation to personal assistance and educational tools.

At the same time, we must recognize that societal and ethical considerations are paramount. ChatGPT, as well as every other AI-based system, will have to strike the right balance between utility and safety.

Ensuring transparency while addressing bias and fairness concerns is a continuous, ongoing challenge that developers and the broader AI community must address as they increase the reliability and convenience of their products.

ChatGPT's journey showcases the incredible progress that has been made in the field of natural language processing. It stands as a symbol of what can be achieved through dedication, research, and a deep commitment to progress. As ChatGPT continues to evolve, we can be sure that it will shape the future of human-AI interaction – even as it contributes to a wide array of applications across industries and disciplines.

Envisioning a Conversational AI-Centric Future

In the not-so-distant future, it is widely expected that conversational AI will become a key component of human-computer interaction. If there is one thing we can be fairly sure of, it's that this transformative technology will permeate our daily lives. Ultimately, it will reshape the way we learn, work, communicate, and entertain ourselves.

Seamless Integration in Daily Tasks

At some point, conversational AI will likely seamlessly integrate into our daily lives. Today, it's already embedded in our computers, smartphones, cars, and even household appliances. In time, we will rely on ChatGPT and similar smartbots for setting reminders, scheduling appointments, ordering groceries, and controlling our smart homes. All these tasks will be handled through natural language conversations.

Personalized Experiences

Conversational AI will eventually become adept at understanding our individual preferences and behavior. We will increasingly rely on it to provide personalized recommendations for everything from travel destinations and fitness routines to movies and books. The more we use it, the more it will adapt and learn, becoming a true digital companion.

In conclusion, I believe it's important to state that this technology is here to stay. Learning to use ChatGPT is not just a convenience, but a necessity for the future of humanity. Resisting change leads only to obsolescence. Embracing it, on the other hand, ensures our relevance in the coming years.

I wrote this book to help prevent people from being intimidated by this exciting new technology, hoping they will instead learn to use it as effectively as possible. So please - go ahead and download the app and usher in a new world of endless possibilities for yourself and your future!

Made in the USA
Las Vegas, NV
18 May 2024

90062930R00069